Third-Class Radiotelephone License Handbook

by

Edward M. Noll

A Revision of

Radio Operators License Handbook

by Edward M. Noll

HOWARD W. SAMS & CO., INC.
THE BOBBS-MERRILL CO., INC.
INDIANAPOLIS · KANSAS CITY · NEW YORK

FOURTH EDITION

SECOND PRINTING—1976

Copyright © 1965, 1971, 1974, and 1976 by Howard W. Sams & Co., Inc., Indianapolis, Indiana 46268. Printed in the United States of America.

All rights reserved. Reproduction or use, without express permission, of editorial or pictorial content, in any manner, is prohibited. No patent liability is assumed with respect to the use of the information contained herein. While every precaution has been taken in the preparation of this book, the publisher assumes no responsibility for errors or omissions. Neither is any liability assumed for damages resulting from the use of the information contained herein.

International Standard Book Number: 0-672-21353-2
Library of Congress Catalog Card Number: 76-19691

Preface

This handbook will serve as a practical study guide for the aspiring radio operator, as well as a ready reference for those working in the field. The laws, rules, regulations, and accepted operating procedures, as they apply to the radio operator, are given in this volume. Information needed by the nonlicensed operator, as well as that needed to obtain the lower-grade FCC licenses, is included.

A knowledge of the internal operation, critical adjustments, and maintenance of radio transmitters and other electronic gear is not required to obtain these lower-grade licenses, so this type of information has been omitted. These licenses are concerned only with the persons who operate the equipment and carry on communications.

Chapter 1 covers the various types of operator licenses and tells when they are necessary. In addition, many helpful operating rules and procedures are given in this chapter. The nonbroadcast services—Maritime, Aviation, Public Safety, Industrial, Land Transportation, and Citizens band—are thoroughly discussed in Chapter 2.

The additional information needed in the operation of broadcast stations is given in Chapter 3. The functions that can be performed by the third-class operator with a broadcast endorsement, the details of log-keeping, and other information that will give you a better technical understanding of station operations are included. Chapter 4 gives the procedures for obtaining a license and the requirements for each classification.

The final three chapters contain questions and answers similar to those given on the actual examination. Chapter 5 covers

Element I, which is concerned with basic law, while Chapter 6 covers Element II, which is concerned with basic operating practices. The test for the third-class license is based on these Elements I and II. To obtain the broadcast endorsement to the third-class license, a test based on Element IX must also be passed; this element is covered in Chapter 7.

Two simple self-tests, patterned after the FCC style of testing, are given. These tests are based on Elements I and II, and Element IX. Answers to the tests are given in Appendix VI.

A knowledge of radio law, including FCC rules and regulations, is essential to pass the examination and be a good operator. Appendices III and IV include those rules and regulations recommended by the FCC for study in preparing for the operator examinations and those you need to know to be a good operator. For example, in Appendix IV you will find the rules and regulations recommended for study by the FCC in preparing for the Element IX examination.

All information needed to secure an FCC license up to, but not including, the second-class license is given in this handbook. Just being able to answer the questions on the exam, however, does not make you an operator. Therefore, much additional information and hints of value to the operator have been included. If the material in this handbook is mastered, you should not only be able to pass the test with flying colors, but be a good operator as well.

EDWARD M. NOLL

Contents

CHAPTER 1

RADIO OPERATORS 7
License Considerations — Essential Provisions for Radio Operators

CHAPTER 2

RADIO SERVICES 21
Maritime Radio Services — Aviation Radio Services — Public Safety Radio Services — Industrial Radio Services — Land Transportation Radio Services — Citizens Radio Service

CHAPTER 3

RADIO BROADCAST OPERATION 69
Technical Considerations — The Operator's Responsibilities — Transmitter Metering — Operating Power — Station Requirements — Log Requirements — Study of Rules and Regulations — Station Plans — FCC Rules and Regulations

CHAPTER 4

LICENSE PROCEDURES AND OPERATOR REQUIREMENTS . . 93
FCC Operator Requirements

CHAPTER 5

ELEMENT I—BASIC LAW 105

CHAPTER 6

ELEMENT II—BASIC OPERATING PRACTICE 109
    General (Series "O") — Maritime (Series "M") — Elements I and II Self-Test

CHAPTER 7

ELEMENT IX—BROADCAST ENDORSEMENT 123
    Element IX Self-Test

APPENDIX I

EXTRACTS FROM THE GENEVA, 1959, TREATY 137

APPENDIX II

EXTRACTS FROM THE COMMUNICATIONS ACT OF 1934, AS AMENDED 139

APPENDIX III

EXTRACTS FROM THE FCC RULES AND REGULATIONS, PARTS 1, 2, 13, 81, AND 83 143

APPENDIX IV

EXTRACTS FROM THE FCC RULES AND REGULATIONS, PARTS 17 AND 73 157

APPENDIX V

INFORMATION CONCERNING COMMERCIAL RADIO OPERATOR LICENSES AND PERMITS 193

APPENDIX VI

ANSWERS TO SELF-TESTS 199

INDEX 203

1

Radio Operators

Radio operators number in the millions in this decade of two-way radio growth. Radio operators are radio operators for a variety of reasons. If you pilot a small boat or fly a small plane, your reason for being a radio operator involves the safety and convenience offered by your two-way radio equipment. Such stations are licensed in the Maritime or Aviation Radio Services, respectively.

Perhaps you are a radio operator because two-way radio expedites your business procedures, as in the Industrial, Citizens, or other land radio services. Your radio operating reason may be a desire for personal-convenience communications (Citizens Radio Service), or a hobby activity (Amateur Radio Service). Your association with radio operation may be as a public service in one of the Public Safety Radio Services.

Some radio operators need not be licensed; others require an FCC license of a particular class. Whether or not you are a licensed operator, you are responsible for the proper operation of the station with which you are associated. It is your duty to know the appropriate laws, rules and regulations, permissible communications, and established procedures for the particular radio service with which you are concerned.

For the most part, those persons involved in radiocommunications have always displayed a certain nobility toward, and respect for, the wonders of radio. Radio waves have been, and are, an instrument of mercy and safety, public service and convenience, basic research and accomplishment, and entertainment and good will. They should not become an instrument of callousness, lawbreaking and selfish exploitation.

LICENSE CONSIDERATIONS

This handbook is concerned with the nonlicensed operator, through the lower-grade FCC licenses, up to, but not including, the second-class radio license. These lower-class licenses are concerned only with the operation of the equipment for carrying on communications. They do not require a knowledge of the technical aspects of two-way radio equipment.

Radio transmitters require a *station license*. The operator of the transmitter and station does, or does not, require an *operator license* depending on the radio service to be rendered. For example, in the Citizens Radio Service a station license is required, but the operator is not licensed. However, the operators of any radio transmitter are responsible to the licensee of the station, and operate with his authorization. Every operator, whether required to be licensed or not, must obey the rules governing the station he operates.

In the aviation and maritime services both *station* and *operator* licenses are required. If you have a station license for a small boat or small plane, you also require a restricted radiotelephone operator permit, as a minimum *operator license*. No examination is required for this license. In other mobile and maritime radio services, higher-grade licenses are required. These licenses involve taking an FCC examination to prove that the applicant possesses the required knowledge.

It is necessary to pass Elements I and II to obtain a *third-class operator permit*. This grade of license is required for those radio services in which it is imperative that the operator know the appropriate rules, regulations, and operating procedures. Such an examination is taken at a district office of the Federal Communications Commission. (For more information regarding license applications and examinations, the reader may refer to Appendix V of this handbook.)

To obtain a *restricted radiotelephone operator permit*, it is only necessary to fill out FCC Form 753. A sample of the form is shown in Fig. 1-1. This application together with a $4.00 fee is mailed to the Federal Communications Commission, Gettysburg, Pa. 17325.

Certain broadcast stations may also be operated with a third-class operators permit provided this license has an appropriate *broadcast endorsement*. Thus, if you wish to operate certain radio broadcast stations, it is necessary that you pass Elements I and II for the third-class license, plus Element IX for the broadcast endorsement. These elements are covered in this license handbook. (Refer to Chapters 5, 6, and 7.)

Specific operator requirements and authority as they appear in the FCC Rules and Regulations, Part 13, are as follows:

(e) *Radiotelephone third-class operator permit:*

(1) Ability to transmit and receive spoken messages in English.

(2) Written examination elements: 1 and 2.

(g) *Restricted radiotelephone operator permit:*

No oral or written examination is required for this permit. In lieu thereof, applicants will be required to certify in writing to a declaration which states that the applicant has need for the requested permit; can receive and transmit spoken messages in English; can keep at least a rough written log in English or in some other language in general use that can be readily translated into English; is familiar with the provisions of treaties, laws, and rules and regulations governing the authority granted under the requested permit; and understands that it is his responsibility to keep currently familiar with all such provisions.

License authorities are as follows:

(g) *Radiotelephone third-class operator permit.* Any station except:

(1) Stations transmitting television other than Instructional Television Fixed Service stations, or

(2) Stations transmitting telegraphy by any type of the Morse Code, or

(3) Any of the various classes of broadcast stations, or

(4) Class I-B coast stations at which the power is authorized to exceed 250 watts carrier power or 1000 watts peak envelope power, or

(5) Class II-B or Class III-B coast stations, other than those in Alaska, at which the power is authorized to exceed 250 watts carrier power or 1000 watts peak envelope power, or

(6) Ship stations or aircraft stations at which the installation is not used solely for telephony, or at which the power is more than 250 watts carrier power or 1000 watts peak envelope power:

Provided, That (1) such operator is prohibited from making any adjustments that may result in improper transmitter operation, and (2) the equipment is so designed that the stability of the frequencies of the transmitter is maintained by the transmitter itself within the limits of tolerance specified by the station license, and none of the operations necessary to be performed during the course of normal rendition of the service of the station may cause off-frequency operation or result in any unauthorized radiation, and (3) any needed adjustments of the transmitter that may affect the proper operation of the sta-

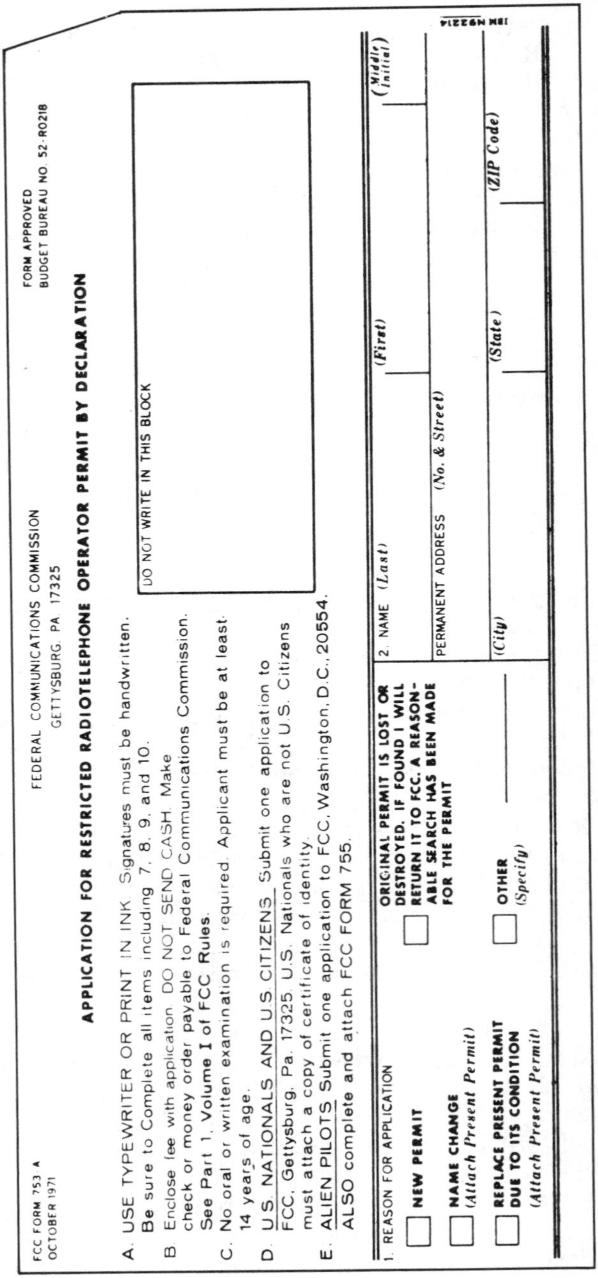

FCC FORM 753 A
OCTOBER 1971

FEDERAL COMMUNICATIONS COMMISSION
GETTYSBURG, PA. 17325

FORM APPROVED
BUDGET BUREAU NO. 52-R0218

APPLICATION FOR RESTRICTED RADIOTELEPHONE OPERATOR PERMIT BY DECLARATION

A. USE TYPEWRITER OR PRINT IN INK. Signatures must be handwritten.
   Be sure to Complete all items including 7, 8, 9, and 10.
B. Enclose fee with application. DO NOT SEND CASH. Make check or money order payable to Federal Communications Commission. See Part 1, Volume I of FCC Rules.
C. No oral or written examination is required. Applicant must be at least 14 years of age.
D. U.S. NATIONALS AND U.S. CITIZENS. Submit one application to FCC, Gettysburg, Pa. 17325. U.S. Nationals who are not U.S. Citizens must attach a copy of certificate of identity.
E. ALIEN PILOTS. Submit one application to FCC, Washington, D.C. 20554. ALSO complete and attach FCC FORM 755.

DO NOT WRITE IN THIS BLOCK

2. NAME (Last) _____ (First) _____ (Middle Initial) _____

PERMANENT ADDRESS (No. & Street) _____

(City) _____ (State) _____ (ZIP Code) _____

1. REASON FOR APPLICATION

☐ NEW PERMIT

☐ NAME CHANGE
   (Attach Present Permit)

☐ REPLACE PRESENT PERMIT
   DUE TO ITS CONDITION
   (Attach Present Permit)

☐ ORIGINAL PERMIT IS LOST OR DESTROYED. IF FOUND I WILL RETURN IT TO FCC. A REASONABLE SEARCH HAS BEEN MADE FOR THE PERMIT

☐ OTHER
   (Specify) _____

Fig. 1-1. A sample of

3. DATE OF BIRTH		PLACE OF BIRTH (City, State, Country)
MONTH	DAY	YEAR	

4. ARE YOU A CITIZEN OF THE U.S.? *(Check one)*   YES ☐   NO ☐

5. Have you been convicted in the last ten years of any crime for which the penalty imposed was a fine of $500 or more, or an imprisonment of more than one year?   YES ☐   NO ☐

(IF YES, FURNISH DETAILS FOR EACH CONVICTION GIVING DATE, NATURE OF CRIME, COURT IN WHICH CONVICTED, NATURE OF SENTENCE AND WHERE SENTENCE WAS SERVED.)

6. DO YOU HAVE ANY OF THE PHYSICAL DEFECTS LISTED BELOW OR ANY OTHER DEFECT WHICH WILL IMPAIR OR HANDICAP YOU IN PROPERLY USING THE PERMIT FOR WHICH APPLYING?   *(Check below, if yes attach details)*

SPEECH IMPEDIMENT	YES ☐	NO ☐
ACUTE DEAFNESS	YES ☐	NO ☐
OTHER	YES ☐	NO ☐

APPLICANT'S CERTIFICATION

I certify that I am the above named applicant; that the facts stated in the foregoing application and all exhibits attached thereto, are true of my own knowledge: that I have need for the permit herein applied for; that I can transmit and receive spoken messages in English; that I can keep at least a rough written station log in English or in some other language in general use that can be readily translated into English; that I am familiar with the provisions of treaties, laws, rules and regulations governing the authority granted under the permit herein applied for; that I understand that it is my responsibility to keep myself currently familiar with all such provisions; that I will preserve the secrecy of radio communications as required by law and that I will faithfully adhere to any requirements of law at all times, that this obligation is taken freely, without mental reservation or purpose of evasion; and that I will well and faithfully discharge the duties of the office obtained through my employment under this Permit if granted.

⑦ _____

(Signature)   (Date)

WILLFUL FALSE STATEMENTS MADE ON THIS FORM ARE PUNISHABLE BY FINE AND IMPRISONMENT. U.S. CODE, TITLE 18, SECTION 1001.

FCC Form 753.

tion are regularly made by or under the immediate supervision and responsibility of a person holding a first- or second-class commercial radio operator license, either radiotelephone or radiotelegraph as may be appropriate for the class of station involved . . . , who shall be responsible for the proper functioning of the station equipment, and (4) in the case of ship radiotelephone or aircraft radiotelephone stations when the power in the antenna of the unmodulated carrier wave is authorized to exceed 100 watts, any needed adjustments of the transmitter that may affect the proper operation of the station are made only by or under the immediate supervision and responsibility of an operator holding a first- or second-class radiotelegraph license, who shall be responsible for the proper functioning of the station equipment.

(h) *Restricted radiotelephone operator permit.* Any station except:

(1) Stations transmitting television, or

(2) Stations transmitting telegraphy by any type of the Morse Code, or

(3) Any of the various classes of broadcast stations other than fm translator and booster stations, or

(4) Ship stations licensed to use telephony at which the power is more than 100 watts carrier power or 400 watts peak envelope power, or

(5) Radio stations provided on board vessels for safety purposes pursuant to statute or treaty, or

(6) Coast stations, other than those in Alaska, while employing a frequency below 30 MHz, or

(7) Coast stations at which the power is authorized to exceed 250 watts, carrier power or 1000 watts peak envelope power;

(8) At a ship radar station the holder of this class of license may not supervise or be responsible for the performance of any adjustments or tests during or coincident with the installation, servicing, or maintenance of the radar equipment while it is radiating energy: *Provided,* That nothing in this subparagraph shall be construed to prevent any person holding such a license from making replacements of fuses or of receiving type tubes: *Provided,* That, with respect to any station which the holder of this class of license may operate, such operator is prohibited from making any adjustments that may result in improper transmitter operation, and the equipment is so designed that the stability of the frequencies of the transmitter is maintained by the transmitter itself within the limits of tolerance specified by the station license, and none of the operations necessary to

be performed during the course of normal rendition of the service of the station may cause off-frequency operation or result in any unauthorized radiation, and any needed adjustments of the transmitter that may affect the proper operation of the station are regularly made by or under the immediate supervision and responsibility of a person holding a first- or second-class commercial radio operator license, either radiotelephone or radiotelegraph, who shall be responsible for the proper functioning of the station equipment.

Additional license information is given in Appendix V.

§ 13.62 Special privileges.

In addition to the operating authority granted under § 13.61, the following special privileges are granted the holders of commercial radio operator licenses:

(b) The holder of any class of radiotelephone operator's license, whose license authorizes him to operate a station while transmitting telephony, may operate the same station when transmitting on the same frequencies, any type of telegraphy under the following condition:

(1) When transmitting telegraphy by automatic means for identification, for testing, or for actuating an automatic selective signaling device, or

(2) When properly serving as a relay station and for that purpose retransmitting by automatic means, solely on frequencies above 50 MHz, the signals of a radiotelegraph station, or

(3) When transmitting telegraphy as an incidental part of a program intended to be received by the general public, either directly or through the intermediary of a relay station or stations.

(c) The holder of a commercial radiotelegraph first- or second-class license, a radiotelephone second-class license, or a radiotelegraph or radiotelephone third-class permit, endorsed for broadcast station operation may operate any class of standard, fm, or educational fm broadcast station except those using directional antenna systems which are required by the station authorizations to maintain ratios of the currents in the elements of the systems within a tolerance which is less than five percent or relative phases within tolerances which are less than three degrees, under the following conditions:

(1) That adjustments of transmitting equipment by such operators, except when under the immediate supervision of a radiotelephone first-class operator (radiotelephone second-class operator for educational fm stations with transmitter output power of 1000 watts or less), and except as provided in paragraph (d) of this section, shall be limited to the following:

(i) Those necessary to turn the transmitter on and off;

(ii) Those necessary to compensate for voltage fluctuations in the primary power supply;

(iii) Those necessary to maintain modulation levels of the transmitter within prescribed limits;

(iv) Those necessary to effect routine changes in operating power which are required by the station authorization;

(v) Those necessary to change between nondirectional and directional or between differing radiation patterns, provided that such changes require only activation of switches and do not involve the manual tuning of the transmitter's final amplifier or antenna phasor equipment. The switching equipment shall be so arranged that the failure of any relay in the directional antenna system to activate properly will cause the emissions of the station to terminate.

(2) The emissions of the station shall be terminated immediately whenever the transmitting system is observed operating beyond the upper and lower limiting values of parameters required to be observed and logged or in any manner inconsistent with the rules or the station authorization, and the above adjustments are ineffective in correcting the condition of improper operation, and a first-class radiotelephone operator is not present.

(3) The special operating authority granted in this section with respect to broadcast stations is subject to the condition that there shall be in employment at the station in accordance with Part 73 of this chapter one or more first-class radiotelephone operators authorized to make or supervise all adjustments, whose primary duty shall be to effect and ensure the proper functioning of the transmitting system. In the case of a noncommercial educational fm broadcast station with authorized transmitter output power of 1000 watts or less, a second-class radiotelephone licensed operator may be employed in lieu of a first-class licensed operator.

(d) When an emergency action condition is declared, a person holding any class of radio operator license or permit who is authorized thereunder to perform limited operation of a standard broadcast station may make any adjustments necessary to effect operation in the emergency broadcast system in accordance with the station's National Defense Emergency Authorization: *Provided,* That the station's responsible first-class radiotelephone operator(s) shall have previously instructed such person in the adjustments to the transmitter which are necessary to accomplish operation in the Emergency Broadcast System.

ESSENTIAL PROVISIONS FOR RADIO OPERATORS

All radio operators, licensed or not, should know and are responsible for knowing the following laws and regulations.

Station License

A radio station, other than one belonging to and operated by the Federal Government, shall not be operated unless it is properly licensed by the Federal Communications Commission and the station license is posted or kept available as specified by the rules governing the particular service and/or class of station.

Operator Licenses or Permits

Except as may be provided otherwise by the rules governing a particular service and/or class of station, a radio station required to be licensed by the Commission shall be operated only by a properly licensed radio operator who has his license or verification card or his permit in his possession or posted in accordance with the Commission's rules governing the particular service in which he is employed. If the license or permit has been sent in to the Commission for replacement, duplicate, etc., a copy of the application for such replacement, duplicate, etc., shall be exhibited in lieu of the license.

The holder of a restricted radiotelephone operator permit as issued to a United States citizen may operate any station in the fixed and mobile services while using radiotelephony, except:

1. Coast stations other than in Alaska, while using a frequency below 30 MHz; or
2. Ship stations licensed to use telephony at which the power is more than 100 watts carrier power or 400 watts peak envelope power.
3. Radio stations provided on board vessels for safety purposes pursuant to statute or treaty.

Coast stations in the above categories which this grade of operator may operate are limited to those at which the power is not authorized to exceed 250 watts carrier or 1000 watts peak envelope power. The transmitting equipment of any station must be so designed that the stability of the operating frequencies is maintained by the transmitter itself within the limits of tolerance specified by the Commission; adjustments to the radio transmitter, which may cause off-frequency operation or result in improper transmitter operation, shall be made only by, or in the presence of, a person holding a first- or second-class operator license, either radiotelephone or radiotelegraph, who shall be responsible for the proper operation of the equipment.

Nature of Communications

Only such communications as are authorized by the rules governing the radio station operated may be transmitted. False calls, false or fraudulent distress signals, superfluous and unidentified communications, and obscene and profane language are specifically prohibited.

Priority of Communications

Distress calls and messages shall have absolute priority over all other communications. Distress calls may be made without regard to interference to other stations, with due consideration, however, being given to any other distress calls or messages which may be transmitted at the same time. Routine operation shall not be resumed until the distress signals and messages have been cleared.

The order of priority for communications in the mobile service shall be as follows:

1. Distress calls, distress messages, and distress traffic.
2. Communications preceded by the urgency signal.
4. Communications relating to radio direction-finding.
5. Communications relating to the navigation and safe movement of aircraft.
6. Communications relating to the navigation, movements, and needs of ships, and weather observation messages destined for an official meteorological service.
7. Government radiotelegrams: Priorite Nations.
8. Government communications for which priority has been requested.
9. Service communications relating to the working of the radiocommunication service or to communications previously exchanged.
10. Government communications other than those shown in 7 and 8 above, and all other communications. (Art. 37.)

Secrecy of Radiocommunications

The contents of a radiocommunication shall not be divulged to any person or party other than to whom it is addressed, except as specifically provided in section 605 of the Communications Act.

Identification of Communications

When not required to identify itself by some other provisions of the Rules and Regulations, every radio station shall identify itself by its regularly designated call signal or other approved method at the time of each transmission, and as frequently as is practicable during tests or during an exchange of long communications.

Prevention of Interference

Inasmuch as most radio transmission in the mobile services is conducted on radio channels which are shared by many stations, as on a "party line," it is necessary that certain precautions be observed to avoid unnecessary congestion and interference.

In order to avoid interference with communications in progress, an operator shall listen on the frequencies on which he intends to receive for a period sufficient to ascertain that he will be able to hear the station he is calling and that his transmission will not cause harmful interference. He shall not attempt to call if interference with established communications is likely to result.

Attempts to establish communication beyond the normal range of installed equipment usually result in unnecessary occupation of the calling frequency. Except in emergencies, such calling should be avoided.

In order to eliminate the need for undue repetition of communications, voice transmission should be made with maximum articulation. It is well to remember that speech is generally rendered almost unintelligible by speaking too close to the microphone, and it is often lost in extraneous noise when the microphone is held at too great a distance.

Radio Log

Radio logs are required to be kept in certain radio services. These logs must be kept by a person having actual knowledge of the facts to be entered who shall also sign the log as prescribed by the Commission. Logs shall be made available on request by authorized Commission representatives. No log or portion thereof shall be erased, obliterated, or willfully destroyed within the period of retention required by the Rules and Regulations. Any necessary correction may be made only by the person originating the entry; he shall strike out the erroneous portion, initial the correction made, and indicate the date of correction.

Notice of Violations

Any licensee who appears to have violated any provision of the Communications Act of 1934, as amended, or of the Rules and Regulations of the Federal Communications Commission, shall be served with a notice calling the facts to his attention and requesting a statement concerning the matter. Within 10 days from receipt of such notice, or such period as may be specified, the licensee shall send a written answer to the Commission field office, or to the monitoring staion originating the official notice. If an answer cannot be sent or an acknowledgment made within such 10 day period by reason of illness or other unavoidable circumstances, ackowledgment and answer shall be made at the earliest practicable date, shall be complete in itself, and shall not be abbreviated by reference to other communications or answers to other notices. If the notice of violation relates to lack of attention to, or improper operation of, the transmitter, the name and license number of the operator shall be given.

Penalties

The general penalty for violation of the Communications Act (first offense) consists of a fine of not more than $10,000 or imprisonment for a term of not more than one year, or both.

The penalty for violation of the Commission's regulations or the international radio regulations consists of a fine of not more than $500 for each and every day during which such offense occurs.

The Commission has authority, as public convenience, interest, or necessity requires, to suspend the license (permit) of any operator on proof sufficient to satisfy the Commission that the operator:

(a) Has violated any provision of any act, treaty, or convention binding on the United States, which the Commission is authorized to administer, or any regulation made by the Commission under any such act, treaty, or convention; or

(b) Has failed to carry out a lawful order of the master or person lawfully in charge of the ship or aircraft on which he is employed; or

(c) Has wilfully damaged or permitted radio apparatus or installations to be damaged; or

(d) Has transmitted superfluous radio communications or

signals or communications containing profane or obscene words, language, or meaning, or has knowingly:
  1. Transmitted false or deceptive signals or communications; or
  2. Transmitted a call signal or letter which has not been assigned by proper authority to the station he is operating; or
(e) Wilfully or maliciously interfered with any other radio communications or signals; or
(f) Obtained or attempted to obtain, or has assisted another to obtain or attempt to obtain, an operator's license by fraudulent means.

Distress Procedure

The international radiotelephone distress signal consists of the spoken expression MAYDAY. This signal shall be used to announce that the ship, aircraft, or other vehicle that sends the distress signal is threatened by serious and imminent danger and requests immediate assistance. The distress signal shall be followed by the distress messages containing the identity of the station in distress, its position, the nature of the distress, and the nature of the assistance requested.

The international radiotelephone urgency signal consists of the word PAN, spoken three times, and is to be used when the calling station has a very urgent message to transmit concerning the safety of the ship, aircraft, or other vehicle, or concerning the safety of some person on board or sighted from on board.

The international radiotelephone safety signal consists of the word SECURITY spoken three times, and announces that the station is about to transmit a message concerning the safety of navigation, or giving important meterological warnings.

The distress *call* sent by radiotelephony comprises:

1. The distress signal MAYDAY spoken three times;
2. The words THIS IS, followed by the identification of the mobile station in distress, the whole repeated three times.

The distress call should be followed as soon as possible by the distress message, which comprises:

1. The distress signal MAYDAY;
2. The name of the ship, aircraft, or vehicle in distress;
3. Particulars of its position, the nature of the distress, and the kind of assistance desired;

4. Any other information which might facilitate the rescue.

As a general rule, a ship shall signal its position in latitude and longitude, using figures for the degrees and minutes, together with one of the words NORTH or SOUTH and one of the words EAST or WEST.

As a general rule, and if time permits, an aircraft shall transmit in its distress message the following information: estimated position and time of the estimate; heading in degrees (state whether magnetic or true); indicated air speed; altitude; type of aircraft; nature of distress, type of assistance desired, and any other information which might facilitate the rescue (including the intention of the person in command, such as forced alighting on the sea or crash landing).

After the transmission by radiotelephony of its distress message, the mobile station may be requested to transmit suitable signals followed by its call sign or other identification, to permit direction-finding stations to determine its position. This request may be repeated at frequent intervals if necessary. In radiotelephony, the signal is made by holding the transmitter on the air for the specified period of time, without speaking into the microphone other than to identify the station.

Immediately before a crash landing or forced landing (on land or sea) of an aircraft, as well as before total abandonment of a ship or an aircraft, the radio apparatus should be set for continuous emission, if considered necessary and circumstances permit.

2

Radio Services

This chapter covers each of the major two-way radio services and categories with particular emphasis on operator license requirements and responsibility. Discussions of typical operations and installations are given. There is some coverage of station license considerations, plus operating frequency bands and important spot frequencies for the various radio services.

MARITIME RADIO SERVICES

The maritime industry was the first enthusiastic and worldwide user of two-way radio. Early in this century radio stations were installed on sea-going vessels. Military, government, and private land stations were set up to maintain contact with these radio-equipped vessels. Obviously, the two major divisions of the maritime service are stations on land and stations on shipboard. There are, of course, a number of subdivisions of both land and shipboard stations.

Two-way radio is compulsory on most vessels. There must be an efficient radio installation in operating condition and in charge of, and operated by, a qualified operator or operators on any ship of the United States (other than a cargo ship of less than 300 gross tons) to be navigated in the open seas outside of a harbor or port. No such vessel may leave or attempt to leave any harbor or port of the United States for a voyage on the open seas unless it is so equipped.

Furthermore, any vessel, regardless of size, that is transporting more than six passengers for hire and is navigated in the open seas or any tidewater within the jurisdiction of the

United States adjacent or contiguous to the open seas must be equipped with an acceptable radio installation. The FCC may exempt from the provision of this part any vessel or class of vessel where the route or condition of the voyage or other condition or circumstances are such as to render a radio installation unreasonable, unnecessary, or ineffective.

Most vessels on which a radio installation is compulsory must include a main radiotelegraph installation and, in most cases, an emergency or reserve radiotelegraph installation. For cargo ships of less than 1600 gross tons, a radiotelephone may be installed in lieu of a radiotelegraph installation.

Three major categories of coast stations are: public, limited, and marine-utility stations. The public and limited stations are further subdivided into three additional categories, Classes I, II, and III. A public coast station is one that is said to be open to public correspondence. Public correspondence itself refers to any telecommunications which the offices and stations must, by reason of their being at the disposal of the public, accept for transmission. When suitable rates are filed, it is possible to charge for such correspondence. A limited coast station is one that is not open to public correspondence but serves mainly the operational and business needs of ships.

Be it public or limited, a Class-I coast station provides a mobile radio service to ships at sea, including such service over distances up to several thousand miles. Class-I coast stations are the only ones that have frequency assignments below 150 kHz or between 5000 and 25,000 kHz. A Class-II coast station provides a maritime service primarily of a regional character. Frequency assignments are not made below 150 kHz or between 5000 and 25,000 kHz. The Class-III coast station provides a maritime service of a local nature and does not operate on any frequency below 25 MHz. The marine-utility station is one that is readily portable for use as a limited coast station at unspecified points ashore within a designated local area.

Two main categories of ship stations are also public and limited. A limited ship station is not open to public correspondence and must confine itself to the operational and business needs of shipping. Public ship stations are also further classified according to their hours of service for telegraphy in a public correspondence service. These categories are based on whether they provide continuous service of public correspondence or a limited service.

The equipment aboard those vessels on which a radio installation is compulsory must meet specific requirements. The installations are inspected and are issued safety certificates attesting to their compliance with the radio requirements of the Safety Convention.

Operator Requirements

In general, the classes of operator licenses required for the compulsory shipboard radio installations are above the license classes which are the major concern of this handbook. From the classes of ship and coast stations covered in the previous paragraphs, it is apparent that most operators must have a radiotelegraph license. Except for certain low-power coast stations, the grade of license must be at least a second-class radiotelegraph grade. For those vessels equipped with a compulsory radiotelephone installation only, a minimum second-class radiotelephone license is needed. Certain of the harbor land-mobile stations which are associated with a larger coast station do not require a licensed operator. However, in this case the communications are under the control of the properly licensed operator of the main station.

One might conclude from the previous discussion that lower license grades have no place in the maritime radio service. However, some ship and coast stations operate with a power of less than 250 watts. There are many types of vessels excluded from the compulsory radio installations. These are the teeming numbers of small commercial and pleasure boats, sailing boats, certain yachts, etc. Two-way radio installations on these boats require a station license and an operators license. The station-license application is made on FCC Form 502; the operator license application is made on Form 753. The operator license is a restricted radiotelephone operators permit, and it is obtained without an examination. The fact that the license does not require an examination does not excuse the operator from knowing the laws, rules and regulations, and operating procedures associated with the particular radio services.

The primary objective of two-way radio in the maritime radio services is safety of life and property. Thus, all operators are responsible for knowing distress procedure. Important reminders concerning FCC rules as they apply to ship radiotelephone and the distress procedure follow.

Ship Radiotelephone Rule Reminders

1. Post station license.
2. Have operator license available.
3. Listen on 2182 kHz.

4. Use 2182 kHz only for calling, distress, urgency, or safety.
5. Listen before transmitting. Avoid interference with distress or other communications in progress.
6. When you hear MAYDAY—listen. Don't talk unless you can help.
7. No ragchewing.
8. Talk 3 minutes, wait 10 minutes.
9. Give your call sign.
10. Keep a log.
11. Answer violation notices.
12. Use of indecent language or profanity on the air is a criminal offense.
13. FALSE OR FRAUDULENT DISTRESS SIGNALS ARE PROHIBITED.

If You Are In Distress

1. Send radiotelephone alarm signal, if possible, to attract attention of other ships.
2. Say slowly and distinctly on the distress frequency of 2182 kHz:
    a. MAYDAY, MAYDAY, MAYDAY
       This is (Call Sign, repeated 3 times)
    b. Give the name of your ship.
    c. Give your geographical position.
    d. Tell the nature of the distress.
    e. Explain what kind of assistance you need.
    f. Give any information that will help you to be rescued. (For example, color of ship, type of ship, length of ship, etc.)
3. Repeat distress call and distress message at intervals until you get an answer.
4. Try any other available frequency to get help, if you get no answer to your distress call sent on 2182 kHz.
5. Give priority to DISTRESS, URGENCY, and SAFETY messages in that order.

Let us consider some of the rules in more detail. The radiotelephone transmitter of a ship station operating on frequencies below 30 MHz may be operated only by a licensed operator. The licensed operator may permit others to speak over the microphone if he starts, supervises, and ends the operation, makes the necessary log entries, and gives the necessary identification.

The license usually held by radio operators aboard small vessels not required to carry a radio installation for safety purposes is the restricted radio operators permit. This is a lifetime permit. However, it does not authorize transmitter adjustments that may affect the proper operation of the station. Any needed adjustments must be made by the holder of a first- or second-class radiotelegraph or radiotelephone license only. It is not necessary to post the restricted radiotelephone-operator permit if it is kept on the operator's person. However, other classes of licenses must be conspicuously posted at the principal location at which the station is operated.

The frequency of 2182 kHz is the calling and distress frequency. Ship radiotelephone stations in the 1600- to 3500-kHz band must maintain an efficient listening watch on this frequency while the station is open and not transmitting on other frequencies. All shipboard transmitters in this band must be capable of transmitting on this frequency, and if the transmitter is used for other than safety communications, it shall also be capable of transmitting on at least two other so-called working frequencies. There are also certain intership frequencies that may be employed. These frequencies are limited to use for safety and operational communications, and in the case of commercial transport vessels they may be used for business communications.

The transmissions from a ship should be confined to the allocated frequencies. However, a ship may transmit on frequencies not incuded on the ship's station license when directed to do so by U. S. Government stations or foreign coast stations.

A number of definite operating procedures must be obeyed. Before transmitting, always listen on the channel to be used so as to minimize interference. You must give your call sign whenever you call another vessel or coast station, and also when you finish the conversation. Except when talking on the intership frequencies where the maximum time limit for conversations is three minutes, you must break and announce your call signs if your ship-to-shore conversation lasts more than 15 minutes. Make your calls short (not more than 30 seconds) and do not call that same station again for two minutes. If a call sign has not as yet been assigned, you may identify your station by announcement of the vessel name and name of the licensee.

If you hear a radio conversation not intended for you, you cannot lawfully use the information in any way. Do not forget that safety is the primary reason for having shipboard radio. Distress and safety must have absolute priority. This is the reason for the setting aside of the distress frequency of 2182 kHz. You transmit on this frequency when you are in distress, and you maintain a watch on this frequency so that you might help another in distress.

It is necessary to keep a radio log. Each page must be numbered, must have the name and call sign of the vessel, and must be signed by the operators. Start and end of the watch on 2182 kHz must be recorded. All distress and alarm signals and related communications transmitted or intercepted and all urgency and safety signals and related communications transmitted shall be recorded in the log as completely as possible.

A record of all installations, service or maintenance work performed, which may affect the proper operation of the station, must also be entered by the licensed operator doing the work, including his signature, address, class of license, serial number, and expiration date of his license. Use the 24-hour system in the radio log; that is 8:45 A.M. is written as 0845 and 1:00 P.M. becomes 1300.

Radio logs must be retained for at least one year—three years if they contain entries concerning distress or disaster. If at any time you receive an official notice of violation from the FCC, you must reply to it within ten days.

Transmitters

Each radiotelephone transmitter used in a ship station must be type accepted under Part 83 of the Commission's Rules. Except for transmitting equipment required to comply with Title III, Part II of the Communications Act, no application for modification of license is required for the deletion, addition, or replacement of radiotelephone and radar transmitters which operate in the frequency bands specified on the license. The additional or replacement transmitters must be type accepted or type approved, as appropriate.

Transmitters for the band 1605-3500 kHz which are installed in a ship station after January 1, 1972 must be capable of single sideband emission, and these stations must also be equipped for transmission in the band 156-158 MHz. These same requirements apply to all ship stations in the band 1600-3500 kHz after January 1, 1977.

2182 Kilohertz

This is the calling and distress frequency for ship radiotelephone stations in the 1605-3500 kHz band, and these stations must maintain an efficient listening watch on 2182 kHz while the station is open and not transmitting on other frequencies. (Rule 83.223.) All ship stations in this band must be capable of transmitting on 2182 kHz.

Recommended Channels for Ship Stations Using Channels in the 1605-3500 Kilohertz Band

Ship radiotelephone stations in the 1605-3500 kHz band if used for other than safety communications shall be capable of transmitting on at least two working frequencies (Rule 83.106(a)). There are eight intership frequencies in the band, as shown in Table 2-1.

Table 2-1. Intership Frequencies

Frequency (kHz)	Geographic area
2003	Great Lakes only.
2082.5	All areas.
2142	Pacific coast area south of latitude 42 degrees north, on a day only basis.
2203	Gulf of Mexico.
2638	All areas.
2670	All areas.
2738	All areas except the Great Lakes and the Gulf of Mexico.
2830	Gulf of Mexico only.

Use of these intership frequencies is limited to safety and operational communications, except that commercial transport vessels may use them also for business communications. (Rule 83.358(a)).

156.8 Megahertz

This is the calling and distress frequency for ship radiotelephone stations in the 156-162 MHz band, and these stations must maintain a watch (Rule 83.224) and be capable of transmission on 156.8 MHz (Rule 83.106), 156.3 MHz and one or more working frequencies.

Recommended Channels for Ship Radiotelephone Stations Using Channels in the 156-158 Megahertz Band

Ship radiotelephone stations in the band 156-158 MHz must be capable of transmitting on 156.8 MHz, 156.3 MHz, and at least one more working frequency (Rule 83.106). Table 2-2 has been prepared as a guide to assist you in deciding what channels to install in your vhf radio, and to determine which channels to use in a particular situation. The table covers both

Table 2-2. Channel Designations

Channel Designators	TYPE OF COMMUNICATIONS Points of communications	Commercial vessels Channel capability						Recreational vessels Channel capability					
		4	6	8	12	16	24	4	6	8	12	16	24
		No. of recommended channels of each group Select channels used in area of operation												
16 (mandatory)	DISTRESS, SAFETY & CALLING Intership & ship to coast	1	1	1	1	1	1	1	1	1	1	1	1
06 (mandatory)	INTERSHIP SAFETY Intership	1	1	1	1	1	1	1	1	1	1	1	1
65, 66, 12, 73, 14, 74, 20	PORT OPERATIONS Intership & ship to coast		1	2	2	3	6		1	1	2	3	7
13	NAVIGATIONAL Intership & ship to coast	1	1	1	1	1				1	1	1	
15 and 162.550 MHz	ENVIRONMENTAL Ship receive only				1	2	2			1	1	2	2
17	STATE CONTROL Ship to coast				1	1	1				1	1	1
07, 09, 10, 11, 18, 19, 79, 80	COMMERCIAL Intership & ship to coast	1	1	1	2	2	4						
67, 08, 77, 88	COMMERCIAL Intership		1	1	1	3							
68, 09	NONCOMMERCIAL Intership & ship to coast							1	1	1	1	1	2
69, 71, 78	NONCOMMERCIAL Ship to coast								1	1	1	1	3
70, 72	NONCOMMERCIAL Intership										1	2	2
24, 84, 25, 85, 26, 86, 27, 87, 28	PUBLIC CORRESPONDENCE Ship to public coast	1	1	1	2	4	5	1	1	2	2	3	5

commercial and recreational usage for radios with 4, 6, 8, 12, 16, or 24 channel capability. The final selection will be determined by comparing this table with the facilities available in your area of interest.

Marine Radiotelephone

Marine radiotelephones operate in the 1605-3500 kHz and 156-158 MHz bands. The distress and calling frequencies are 156.8 MHz and 2182 kHz. A ship-to-shore radiotelephone as installed in a pleasure boat is shown in Fig. 2-1. The mounting position is convenient to the helmsman. To the left of the two-way radio is a radio direction finder. Such a unit can be used to take bearings on low-frequency shore-beacon stations and a-m radio-broadcast stations.

Over the next several years many equipment and band distribution changes will occur. Short-range communications will occur more frequently on the vhf rather than mf bands. The mf band will change over to single-sideband transmission (SSB) rather than conventional DSB amplitude modulation.

SSB requires only one-half the emission bandwidth of standard a-m modulation, conserving spectrum space. Furthermore SSB licenses will be granted only on the basis that the ship station also includes vhf facility. Fm is the dominant form of modulation on the vhf band. Change-over scheduling is as follows:

Courtesy Raytheon Co.
Fig. 2-1. Ship-to-shore radiotelephone.

No new DSB (conventional a-m) ship stations may be installed after January 1, 1972. However, where authorized prior to January 1, 1972, DSB transmitters may continue to be authorized *to the same licensee* until January 1, 1977. (January 1, 1977 is the date for discontinuance of DSB aboard all vessels.) The expression "to the same licensee" means that if an owner of a boat equipped with a DSB transmitter sells his boat during that interim period of January, 1972 to January, 1977, the *radio is not re-licensable*. However, the owner can transfer a DSB set from an old boat to a new one in his own name and a new license would be granted for the period up to January 1, 1977; in this case a vhf radio would not be required.

Coast Stations will discontinue the use of DSB on January 1, 1972; however they will continue to receive DSB until January 1, 1977. After January 1, 1972 Coast Stations will have capability of transmitting A3H (full carrier) so that operation with DSB Ship Stations will not be impaired before the date of January 1, 1977 (the date of discontinuance of all DSB).

After January 1, 1972, any new installation in the 1600-4000 kHz range will be SSB and only Ship Stations that already have vhf will be granted authorization for SSB. This is being done to force all short range communications to vhf.

After January 1, 1977, 2 MHz frequencies will be available only to Public and limited Coast Stations where service is also provided on vhf.

After January 1, 1977 Public Coast Stations serving lakes or rivers will not be authorized to use frequencies in the band 2000-2850 kHz (except on the Mississippi River system and Great Lakes).

Examples of modern medium-frequency and vhf transceivers are shown in Fig. 2-2. The medium-frequency unit (Fig. 2-2A) is a single-sideband model with a 100 watt peak-envelope power (PEP) rating. The single-sideband circuits are such that either the upper or the lower sideband can be

Courtesy Sideband Associates

(A) Single-sideband medium-frequency marine transceiver.

Courtesy Pearce-Simpson

(B) A vhf marine transceiver.

Fig. 2-2. Two modern transceivers.

transmitted. Usually the lower sideband is employed on the medium frequency marine band. This model has facilities for six channels.

Single-sideband emission is specified as A3J emission by the FCC. A unique feature of the transceiver is that it can also transmit what is known as A3H emission which involves the transmission of a carrier and one sideband. The A3J emission is strictly a single-sideband form with no carrier. The A3H capability permits the transmitted signal to be received by a standard a-m receiver. This mode of transmission will be useful during the several years of changeover between standard a-m modulation and strictly single-sideband modulation.

The transceiver in Fig. 2-2B is for vhf-band operation using frequency modulation. There are 12 crystal-controlled transmit frequencies and 14 crystal-controlled receive frequencies. These are selected according to usage and channel designator system.

It has a number of useful facilities. On high power it has a 25-watt capability; on low power, a 1-watt rating. Thus for communications with a nearby station one can use very low power so as not to cause interference to more distant stations. The transmitter can be operated over a frequency range of 156.275 to 165.4245 MHz. The receiver operates over a range of 156.275 to 163.275 MHz, permitting the reception of key coast stations.

Both models shown in Fig. 2-2 are entirely solid state except for the final power stages of the transmitter.

The unit shown in Fig. 2-3 is a piece of navigation equipment. The unit operates on the aircraft-frequency spectrum

Courtesy Triton Marine Products

Fig. 2-3. Direction finder for taking bearings on aircraft beacon stations.

between 108 and 118 MHz. There are many aircraft-beacon stations (VOR) along the coasts that transmit in this range. The OMNI direction finder can be used to obtain a very accurate bearing by tuning in one or several of these aeronautical beacon stations.

AVIATION RADIO SERVICES

The Aviation Radio Services set aside portions of the spectrum for radiocommunication and radionavigation facilities for aircraft operators, aeronautical enterprises, and organizations that require radio-transmitting facilities for safety purposes and other necessities. The radio stations are allocated on the basis of a number of categories—airborne-aeronautical advisory, aeronautical multicom, aeronautical enroute, aeronautical metropolitan, flight test, flying school, airdrome, aeronautical utility, aeronautical search and rescue mobile, aeronautical fixed, operational, radionavigation land, and civil air patrol.

Most of these stations require a licensed operator. Again, insofar as installation and maintenance are concerned, a higher grade license is needed, first- or second-class radiotelephone. If radiotelegraphy is employed, the operator must have at least a third-class radiotelegraph license. Certain stations require a second-class radiotelegraph license.

In general, communications in the aviation services shall be restricted to safe, expeditious, and economical operation of aircraft, and the protection of life and property in the air. Some of the Aviation Radio Services, namely aeronautical public service, aeronautical advisory, aeronautical multicom stations, and civil air patrol land and mobile stations may conduct additional communications in accordance with the particular service of which they are a part.

As in other radio services, each transmitter must be licensed. Except for certain land mobile units used at airdromes and for other aeronautical applications, each station must be operated by a licensed operator. Quite often only a *restricted radiotelephone operator permit* is needed. Although no examination is required in obtaining a permit, as mentioned in Chapter 1, the operator is responsible for knowing the laws and FCC rules and regulations that apply to the particular radio service with which he is associated.

In summary, application must be made for both station and operator licenses for airborne and ground aeronautical radio stations. FCC Forms 404, 406 and 753 are used.

Operator Requirements

Practically all airborne and aeronuatical ground radio stations require that the operators be licensed. In most cases the license is only a restricted radiotelephone permit. As covered previously, no examination is required, and the license is issued for life. For example, such a license is required if you fly a private aircraft that employs two-way radio. Such an operator license is also required for the operator of a two-way radio used in a small private airfield.

Again, although the operator holds only a restricted permit, he must be familiar with the appropriate operating procedures and rules that apply to the particular radio station. Installation, maintenance, and any adjustments that influence the radiation from the transmitter must, of course, be made by an operator with a second-class radiotelephone license or higher. If radiotelegraphy or other form of coded transmission is used, the operator must have an appropriate radiotelegraph license or permit.

For higher-powered stations and more complex equipment using directional antennas and other facilities, higher grade licenses are often required. Usually for transmitters with authorized powers in excess of 250 watts carrier or 1000 watts peak envelope power, the minimum grade license is a second-class type.

Station Categories

There are a variety of airborne and ground aeronautical stations. The four classifications of airborne stations are: air carrier, private, flight test and flying school, and aeronautical public service. An air-carrier aircraft station is one aboard an aircraft engaged in, or essential to, the transportation of passengers or cargo for hire. A private aircraft station is one aboard an aircraft not operated as an air carrier.

An exception to the allocation of air carrier or private aircraft stations involves the weight of the aircraft. If it is less than 12,500 pounds, it may be considered, at the option of the applicant, as a private aircraft even though it is actually engaged in air-carrier operations.

Obviously, station licenses are also granted to aircraft used at test facilities or in flying schools. Licenses can also be obtained for aircraft used for the handling of public correspondence in the same manner as such services are available to ships in the Maritime Radio Service. These stations can handle messages for hire. Facilities must be made available for the use of all persons and without discrimination; such stations shall intercommunicate with any other stations similarly licensed when necessary for the handling of traffic.

Aircraft radio stations using radiotelephony, with the exceptions noted in § 87.133 of the FCC rules, shall be operated by persons holding any class of commercial radio operator license or permit. All transmitter adjustments or tests during, or coincident with, the installation, servicing, or maintenance of a radio station that may affect the proper operation of such station, shall be made by or under the immediate supervision or responsibility of a person holding a first- or second-class commercial radio operator license, either radiotelephone or radiotelegraph, who shall be responsible for the proper functioning of the station equipment.

There is even a greater variety of aeronautical ground stations. Airdrome control stations provide communications limited to the necessities of safe and expeditious operation of aircraft using the airdrome facilities or operating within the airdrome control area. Such stations are required to monitor the following frequencies during hours of operation—121.5 MHz and 3023.5 kHz (Alaska only).

In association with the airdrome there are also certain land mobile stations called aeronautical utility and aeronautical search and rescue mobile stations. Such stations can be installed in fuel trucks, airdrome repair trucks, emergency vehicles, etc.

Throughout the United States there are numerous radio-navigation land stations that guide and give instructions to aircraft enroute. Most of these stations are operated by the Federal Aviation Agency. These stations along with the airdrome radio facilities provide radionavigation coverage that extends from destination to destination, both on and off the civil air routes.

For this fundamental purpose there are certain other station classifications operated by the FAA, or through private ownership if the applicant justifies the need for the facility and the government is not prepared to render this service. These stations are aeronautical enroute, aeronautical metropolitan, aeronautical fixed, and operational stations.

Small airports and landing areas can be allocated aeronautical-advisory and aeronautical-multicom station licenses. When operated at a landing area not served by an airdrome control station, communications must be limited to the necessities of safe and expeditious operation of private aircraft, pertaining to the condition of runways, types of fuel available, wind conditions, weather information, dispatching, or other necessary information. Also, on a secondary basis, information concerning ground transportation and food or lodging can also be handled.

The same information can be communicated via an aeronautical advisory station in an area served by an airdrome control station, with the exception that information concerned with runway conditions, fuel, and weather are handled by the airdrome control station.

A multicom station can be used to direct activities associated with such aeronautical activities as fighting forest fires, aerial advertising, parachute jumping, etc.

Ground base facilities can also be provided for flying schools, flight-test facilities, and Civil Air Patrol.

All of the various aeronautical ground stations just covered must be operated by FCC-licensed operators, with the exception of the FAA stations operated by government employees. Also, certain land mobile stations do not require an operator; usually such vehicles are associated with an airdrome control station or other aeronautical ground station which is under the control of a licensed operator.

Aeronautical Radiocommunication and Radionavigation

There are many electronic aids to aircraft navigation; they can be considered in the four general categories of communication, navigation, traffic control, and landing. Most of the modern-day aeronautical radio activity occurs in the frequency spectrum between 108 and 136 MHz. Radionavigation uses the spectrum between 108 and 118 MHz; air traffic control, 118 to 136 MHz. Spotted throughout this spectrum are frequencies assigned to both aircraft and aeronautical ground stations. For example, in flying a private aircraft you will find the frequencies of the various ground stations given on navigation maps and/or charts. These aeronautical ground stations monitor certain aircraft frequencies and you can quickly establish contact enroute by setting your aviation radio to an appropriate frequency.

Many aviation radio units have dual-reception facilities. Thus it is possible to receive a radio navigation signal continuously at the same time that a two-way radio contact is being made with an aeronautical ground station. Such a unit is often referred to as a one and one-half communicator because it has a single transmitter and two receivers.

Most modern flying is done via the vhf Omnidirectional Range stations. These are called VOR or OMNI stations. In the vhf frequency spectrum there is largely static-free reception, and the bending and false beams of the older low-frequency radio range stations are not present. A reliable directional pattern can be produced at these frequencies. A complex revolving antenna pattern that uses electronic switching generates a rotating beam that has a directional accuracy of 1° throughout the entire 360° of rotation.

A reasonably complete Narco installation in a Piper Cherokee Arrow is shown in Fig. 2-4. The two vertical rows of basic equipment begin immediately to the right of center on the air-

Courtesy National Aeronautical Corp.

Fig. 2-4. Aircraft radio installation.

craft control panel. The radio and navigation switcher is located at the top of the first panel row, followed by two communicators. Each communicator contains a transmitter and receiver that operate on the frequency indicated by its digital readout. Thus each communicator can be used as a simplex transceiver operating on the same frequency. However, duplex operation is possible by setting one communicator to the desired transmit frequency and the second communicator to a designated receive frequency. In this manner of operation the transmit and receive frequencies differ.

The navigation receiver is located immediately below the two communicators. Associated meters indicate bearing and right-left positioning of aircraft with respect to the VOR range station being received by the navigation receiver.

A transponder is the last unit of this row. When a transponder is activated it sends out a coded signal. At a ground station, aircraft can be seen and identified on a radar screen. A very strong signal is received because the transponder is activated whenever the aircraft is searched for by the strong beam from the ground-based radar station. Furthermore the return signal is a coded one and this information is also displayed. Therefore the ground station is not only able to determine the distance and direction to the plane but also is able to make a positive identification of the particular plane because of its returned transponder signal.

At the top of the second row is a radio direction finder. This receiver can be tuned to the low frequency range station. Its associated meter (top and immediately to the left of the first panel of equipment) shows the relative bearing of the range station with respect to the aircraft heading.

The bottom unit of the second panel is a distance-measuring (DME) unit. In operation it sends out an interrogating signal to a special range facility called a Vortac station. The Vortac station, in turn, sends back a signal to the aircraft unit. By utilizing the time of travel of the interrogation signal and the return signal, it is possible to determine the distance to the Vortac station. The actual range in nautical miles is indicated by the meter on the left side of the instrument. A groundspeed indicator, the meter at the lower left of the panel, operates in conjunction with the distance-measuring unit. It reads ground speed in flying toward or away from the Vortac station.

Emergency and Distress

121.5 MHz is a universal simplex clear channel for use by aircraft in distress or condition of emergency. It will not be assigned to aircraft unless other frequencies are assigned and available for normal communications needs. The channel is available, as follows:

1. For emergency communications when circumstances beyond the control of the pilot prevent communication between the aircraft and ground stations on other regularly assigned channels.
2. For emergency direction finding purposes.
3. For establishing air-to-ground contact by aircraft in distress, emergency or when lost.
4. In connection with search and rescue operations, to provide a common channel for aircraft (either civil or military) not equipped to transmit on 123.1 MHz. This includes communications between aircraft, and between

aircraft and ground stations. Stations having the capability should change to 123.1 MHz as soon as practicable.
5. To provide a common frequency for survival communications and for survival radio beacons (emission A2).
6. For air/ground communications between aircraft and ocean station vessels for safety purposes when service on other vhf channels is not available.

The frequency 243 MHz is available to survival craft stations which are also equipped to transmit on 121.5 MHz.

Equipment Tests

Aircraft stations are authorized to make routine tests when required for the proper maintenance of the station, provided that precautions are taken to avoid interference with any other station. A call from an aircraft in flight on the frequency 121.5 MHz for the purpose of making an unannounced or unanticipated test of the alertness of a ground station is not permitted.

Station Operation

Private and air carrier aircraft radio stations are generally limited to communications relating to the necessities of safe aircraft operation. Aeronautical public service stations may be authorized for use aboard aircraft to provide a means of conducting public correspondence. The licensee of a radio station is responsible at all times for the proper operation of his station. Thousands of other aircraft stations use the same frequencies that are assigned to your station. Be brief; transmit only essential messages. Your calls will receive quicker response, repeats will be fewer, and a general improvement in aircraft safety communications will result if the following precautions are taken:

1. Shorten or eliminate test calls while on the ramp or in flight.
2. Be sure the channel is clear before transmitting.
3. Tune your receiver to the correct receiving channel before transmitting.

Station Identification

Aircraft stations frequently cause confusion by failure to properly identify themselves when calling or working by radiotelephone. Aircraft radio stations are normally identified by use of the FAA registration number. After the first communication of each series, the last two characters of the registration number may be used if the practice is initiated by the ground station operator.

PUBLIC SAFETY RADIO SERVICES

The Public Safety Radio Services provide a service of radiocommunication essential to the discharge of either nonfederal governmental functions or the alleviation of an emergency endangering life or property. FCC licensed as well as nonlicensed operators and dispatchers are required for these services.

Specific radio station assignments are made under the categories of local government, police, fire, special emergency, highway, forestry-conservation, and state guard. It is to be noted that many of these radio services are closely related. Hence they are suited to the setting up of radio control centers. For maximum benefit and minimum interference and confusion, there should be close cooperation in the selection of operating frequencies.

The radio operator in the Public Safety Service must be a good operator and develop a good understanding of police and emergency procedures. The Public Safety Radio Services are rather closely knit in a manner similar to the individual services that come under both the marine and aviation radio services. For example, a police radio operator or dispatcher should also be familiar with fire and special-emergency radio procedures. The Industrial and Land-Transportation radio categories encompass more divergent fields of interest.

In summary, the public service radio operator is much more closely linked with the general public in comparison to the get-a-job-done objectives of the Land Transportation and Industrial Radio Services, or the safe-journey objectives of an aircraft or boat station in the aviation or marine radio services. The public safety radio operator and dispatcher is a responsible public servant.

Operator Requirements

Public safety radio stations can be operated by unlicensed persons when so authorized by the station licensee. Such operators may be stationed at either control or dispatch points.

For each of the radio services a station license must be obtained. Application is made on FCC Form 400. The exact FCC regulations are as follows:

§ 89.163   Operator requirements.

(a) Operation during the course of normal rendition of service—radiotelephone.

(1) The following classes of stations may be operated by an unlicensed person, if authorized to do so by station licensee:
    (i) From a control point—a mobile, a base or fixed station.
    (ii) From a dispatch point—a base or fixed station.

*   *   *   *   *   *   *

(e) Licensed operator required. Notwithstanding any other provisions of this section, unless the transmitter is so designed that none of the operations necessary to be performed during the course of normal rendition of service may cause off-frequency operation or result in any unauthorized radiation, and unless the transmitter is so installed that all controls which may cause improper operation or radiation are not readily accessible to the person operating the transmitter, such transmitter shall be operated by a person holding a first- or second-class commercial radio operator license, either radiotelephone or radiotelegraph as may be appropriate for the type of emission being used, issued by the commission.

Frequency Advisory

The almost continuous activity of the Public Safety Radio Services requires wise conservation of frequency spectrum and effective operating procedures. Coded signals and special fill-in form transmissions are prevalent to conserve air time and expedite safety activities quickly and effectively. Frequency advisory committees are especially necessary to permit the wise allocation of frequencies and minimum interference among stations that share the same frequency or operate on nearby channels.

An operating-procedure manual has been published under the direction of a national organization, The Associated Public-Safety Communications Officers, Inc. (APCO). This organization acts in a frequency-advising capacity and has done much to consolidate and standardize law enforcement methods and operating procedures as they involve communications.

Radio-Control Centers

The need and growth of public safety radio operations has forced the development of large urban and county radio-control centers. The radio-control center of the police radio system of Bucks County, Pennsylvania, meets the requirements of an expanding suburbia (Fig. 2-5). It is an accelerated growth area that was a rural county only a few years ago. Public Safety Radio Services have been coordinated into a radio-control center, which includes both police and fire radio-control points

Fig. 2-5. Police radio-control center in Bucks County, Pa.

(Fig. 2-6). Civil defense and other special emergency radio services are housed at the same location. Emergency power is available.

The surrounding townships and municipalities are a part of the same police radio system that utilizes four vhf frequency assignments as major outlets. Interzone and coordination communication facilities are also available. The radio-center staff includes a licensed communications director and technician, plus a chief dispatcher and many radio dispatchers. Twenty-four hour service is provided for most of the county's full-time police agencies, and for practically all of the part-time departments.

Such radio-control centers provide a high order of efficiency and effective coordination. The shared and cooperative use of

Fig. 2-6. Fire-control position at radio-control center in Bucks County, Pa.

frequencies reduces interference and speeds message handling. Coordination with other public-safety radio services reduces confusion and permits smooth handling of disaster activities.

A very unusual emergency radio-operating center has been constructed at Rockville in Montgomery County, Maryland (Fig. 2-7). It is a solid well-built installation protected from adverse weather, heat, shock, nuclear radiation, etc. Facilities can house a staff of 75 persons for thirty days on stocked supplies. Power is made available by its own generating plant. Police, fire, and rescue radio systems are a part of the installation, including their respective dispatching rooms. The radio systems are in operation continually on a county-wide basis.

Fig. 2-7. Montgomery County, Md., emergency operating center.

State police, because of the larger land area under their jurisdiction, must rely on two-way radio for rather long-distance liaison as well as more localized communications. Base stations (Fig. 2-8) are strategically located around the state in a manner that can facilitate local and interzone communications. Some radio-relay facilities are also in order along turnpikes and express highways. Elaborate systems require maintenance facilities. A representative test bench and array of test equipment are shown in Fig. 2-9.

INDUSTRIAL RADIO SERVICES

In the Industrial Radio Services part of the radio spectrum has been set aside for radiocommunication and control facilities for various industrial enterprises which, for safety purposes or other necessity, need radio-transmitting facilities in order to function efficiently. These radio services may not be

Fig. 2-8. Pennsylvania State Police base station and emergency generator.

Fig. 2-9. Pennsylvania State Police two-way-radio maintenance facility.

used for a common-carrier service or to carry program material of any kind that will be used in any way in connection with radio broadcasting.

The radio stations are allocated on the basis of a number of categories—power, petroleum, forest, motion picture, relay press, special industrial, business, industrial radio location, manufacturers, and telephone-maintenance radio services. The four general classifications of stations are: mobile-base, fixed, mobile-relay, and fixed-relay stations. The mobile service, as in the radio services discussed previously, involves communications between mobile and base stations or among mobile stations. A fixed-station radio service is a service of radiocommunications among specified fixed locations. A fixed relay station is established to receive radio signals directed to it from any source, and to retransmit them automatically for reception at one or more fixed locations. A mobile relay station is, in effect, a base station in the mobile service which is authorized primarily to retransmit automatically, on a momile service frequency, communications originated by mobile stations.

The functional block diagram of Fig. 2-10 demonstrates the purpose of the four major industrial station types. Notice that the fixed relay station is a retransmission point located between two fixed stations. However, the mobile relay station is a permanent station that is used to provide a service of retransmission for a given mobile system where some of the mobile units must be operated at a further distance than the basic coverage area of the installation, or on the opposite side of an obstruction to radio propagation.

In the industrial radio services, permissible communications are those considered essential to the efficient conduct of that portion of the enterprise for which the licensee is eligible to hold a station license, subject to the condition that harmful interference is not caused to safety communications of stations licensed in these services.

Control of a fixed relay station or a remotely located base station can be handled by a master controller (Fig. 2-11) positioned at a center of activity for a business or industrial complex. Using tone signalling over a telephone line as many as 12 control functions can be handled. The unit includes microphone, speaker, PTT bar, volume control, audio meter, digital clock and selector switches for the function provided.

Permissible communications are also those related directly to the safety of life or the protection of property. In fact, a station licensee under this part may communicate with other

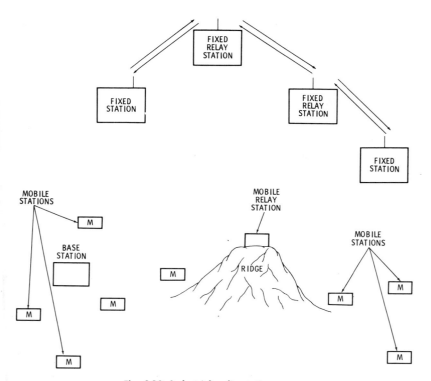

Fig. 2-10. Industrial radio station types.

stations without restrictions as to type, service, or licensee, when the communication to be transmitted involves safety of life or the protection of property.

In the Industrial Radio Services, frequency assignments are allocated within the following bands: 1.6 to 6 MHz, 25 to 50

Courtesy General Electric Co.
Fig. 2-11. A master controller.

MHz, 152 to 174 MHz, and 450 to 460 MHz. For some of the services it is only necessary to make application using FCC Form 400. In other services, the form must be submitted together with evidence of frequency coordination. Frequency coordination, when required, must consider all stations operating on the requested frequency within 75 miles of the proposed station, and all stations operating on any adjacent frequency within 15 kHz of the requested frequency and within 35 miles.

Most stations are allocated for what is referred to as simplex operation. In simplex operation (Fig. 2-12), mobile and base

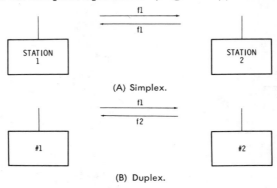

Fig. 2-12. Simplex and duplex operation.

stations operate on the same frequency. In a conversation only one station can transmit at a time. Usually this is referred to as push-to-talk operation, where a given mobile or base station switches between transmit and receive with the use of a microphone switch. In some of the Industrial Radio Services full duplex operation is permitted. In this mode of transmission more than one frequency is used and the radiocommunication can be telephonelike with both transmitters and receivers operating continuously during the communications.

Operator Requirements

During the course of normal rendition of service by radiotelephone, an unlicensed person, if authorized by the station licensee, may operate a base or fixed station from a dispatch point, or may operate a mobile, base, or fixed station from a control point. (If normal operation requires adjustments that could cause improper operation, or if the controls for such adjustments are readily accessible to the operator, then the operator must hold a first- or second-class commercial radio operator license. Also, the use of unlicensed operators does not relieve the station licensee of the responsibility for the proper operation of the station.) Only a person holding a commercial radiotelegraph license or permit (of any class) shall operate a station during the course of normal rendition of service when transmitting radiotelegraphy by any type of the Morse code.

Again, it is important to understand that all transmitter adjustments or tests during or coincident with the installation, servicing, or maintenance of a radio station, which may affect the proper operation of such station, shall be made by or under the immediate supervision and responsibility of a person holding a first- or second-class commercial radio license (must be radiotelegraph license for radiotelegraph stations transmitting by any type of Morse code).

The control point is the key location of a two-way radio installation. It is this point at which the person immediately responsible for the operation of the transmitter is located, and appropriate monitoring facilities are installed. This position must be under the control and supervision of the licensee.

The monitoring facility at the control point must include a carrier-operated device which will provide a visual indication when the transmitter is radiating or a pilot lamp or meter which will provide continual visual indication when the transmitter control circuits have been placed in a condition to produce radiation. At the control point the person responsible for the operation of the transmitter must be able to aurally monitor all transmissions originating at dispatch points. Facilities must be included which will permit the person responsible for the operation of the transmitter to disconnect dispatch-point circuits, and to turn the transmitter carrier on and off at will.

Station identification is also the control operator's responsibility. The assigned call letters should be transmitted at the end of each transmission or exchange of transmissions, or once each 15 minutes of the operating period, whichever the licensee may prefer. Mobile stations may use simple unit identifiers which must be kept on file in the station records of the associated base station.

Station Categories

Three of the categories of the Industrial Radio Services have to do with natural products; these are the power, petroleum, and forest-product services. Power-radio allocations are made to persons engaged in the generation, transmission, or distribution of electrical energy or to persons engaged in the distribution of manufactured or natural gas by means of a pipeline. The service also includes those engaged in the distribution of water or steam. It also applies to the activities associated with the collection, transmission, storage, or purification of water or the generation of steam preparatory to such distribution. The petroleum radio allocations are made to those engaged in prospecting for, producing, collecting, refining, or transporting by means of pipeline, petroleum or petroleum products.

It is apparent that in these systems one can anticipate the use of a number of fixed radio stations, because of the great distances over which these products are transported by pipeline. Of course, base-mobile installations are essential at many locations, such as main distribution centers, prospecting sites, etc. Allocations are available for those persons engaged in tree logging, tree farming, or related woodland operations.

In a large industry a subsidiary corporation can be set up to handle radiocommunications on a nonprofit basis for the parent company or other subsidiaries of the parent corporation. It is also permissible to set up a separate nonprofit corporation or association to handle radiocommunications for persons engaged in one or more of the acceptable activities of the various radio services. Cooperative arrangements are also acceptable when made between two or more persons for the use of radio station facilities. All such persons must be eligible to hold a station license in one of the Industrial Radio Services. It is possible then to share the use of a base station which is licensed to one member of the group. Fixed stations can be operated in the same manner.

In the industrial radio category there are three rather closely related industrial and business services. These are: special industrial, manufacturers, and business radio services. In the manufacturers category the manufacturing activity should include the mechanical or chemical transformation of organic or inorganic substances into new products within establishments usually described as plants, factories, shipyards, or mills, and which employ in that process, power-driven machines and material-handling equipment. The radio service also includes those establishments engaged in assembling components of manufactured products in plants, factories, shipyards, or mills.

The manufacturers radio service does not apply to establishments primarily engaged in the wholesale or retail trade, or in service activities even though they fabricate any or all of the products or commodities handled. These activities are appropriate to the business radio service, which includes any person engaged in a commercial activity. The business radio service also applies to professional people; educational, philanthropic, or ecclesiastical institutions; and hospitals, clinics, and other medical associations.

Still another category is the special-industrial radio service. Allocations are set aside for persons regularly engaged in the operation of farms, ranches, or similar land installations, or for persons engaged in the operation of mines. Special industrial also applies to commercial business operations engaged in the construction of roads, bridges, sewers, pipelines, airfields, etc. Special-industrial allocations are also made to those engaged in specialized services essential to industrial operations or public health, such as soil conservation, seeding, fertilizing, spraying, etc. This service also extends to such varied activities as patroling and repairing of gas and liquid transmission pipelines, water disposal systems, distribution systems of public utilities, cementing, logging, supplying of materials and services to large industrial organizations, delivery of ice or fuel, delivery and pouring of ready-mixed concrete or hot asphalt, etc.

It is apparent that almost any industrial activity is included somewhere in the list of permissible uses for the various industrial radio services. The business radio service has become increasingly popular because of the great help of two-way radiocommunications in professional and service activities. Many wholesale or retail businesses use two-way radio to great advantage. Modern equipment is compact, efficient, and

Courtesy Kaar Engineering

Fig. 2-13. Business radiotelephone installation.

very reliable. The microphone and control unit mount conveniently below the dashboard of a car, as shown in Fig. 2-13. Battery-powered hand-held units (Fig. 2-14) can be used by a person who must move about a large area on foot. Combination mobile and hand-carried units (Fig. 2-15) are also appropriate for some business activities.

Fig. 2-14. A hand-held two-way radio.

Courtesy General Electric Co.

For most activities in the Industrial Radio Services, the various pieces of equipment can be operated by unlicensed persons. Of course, any such operation is only legal when it is authorized by the licensee who is responsible for the correct operation of the radio station. Just as you are responsible for knowing the law pertinent to the operation of an automobile, you are also responsible for knowing the law attendant to the operation of a radio station that may be a part of the very same automobile.

Two other categories in the Industrial Radio Services involve the news and entertainment media of motion picture and press. Allocations in the motion-picture radio service are made to those engaged in the production or filming of motion pictures. Relay press allocations are made to persons engaged in the publication of a newspaper or any operation of an established press association.

There is also a telephone maintenance category for those engaged in rendering a wire-line or wire-line and radiocommunications service to the public for hire.

In Pennsylvania a telephone-maintenance radio service is operated by the privately owned Denver and Ephrata Tele-

Courtesy General Electric Co.

Fig. 2-15. Combination portable/mobile unit.

phone Company (Fig. 2-16). The base-station transmitter is unattended and is located on a ridge not too far from the central office. This location provides a site of high average terrain relative to the service area of the telephone company. The transmitter is remotely controlled from the central office. Automatic switching and monitoring facilities are installed at the central office. From the central office, digital, ringing, and voice signals are switched and routed among base transmitter, mobile units, and regular land telephones.

Mobile units are installed in the company's service trucks and cars. The base-station receiver is tuned to the mobile frequency. It picks up the digital information sent out from the mobile transmitter and routes it over wire line to the

Fig. 2-16. Base-station transmitter of the Denver and Ephrata Telephone Company.

central office. The central office rings the appropriate phone and establishes contact. The circuit is now completed over a land line to the base transmitter; the base transmitter goes on the air and completes the link to the mobile receiver.

By proper selection of mobile frequencies, full duplex operation is possible between a mobile and any regular land telephone. Full duplex operation is also possible between two mobile units.

The control panel of the mobile unit is shown in Fig. 2-17. The key switch has operate and standby positions. On standby

Fig. 2-17. Control panel of a mobile radiotelephone.

the unit rests and waits for an incoming call. In addition to a ring, a green light comes on whenever the particular mobile unit has been called. Whenever the radio system is busy, the amber light comes on. A red light indicates the mobile transmitter is in operation and on the air. The system is planned for five-channel operation; these channels are selected by the push buttons at the top of the panel.

IMTS Telephone System

An effective dial mobile telephone system is in operation. This is referred to as the improved mobile telephone system (IMTS). For this service there are eleven channels allocated in the 150-MHz band. When such a telephone is installed in your car or other land vehicle, you will be able to dial any telephone in your area without contacting a mobile service operator. Furthermore, in traveling into other cities you will be able to dial into their radio system and make calls to any telephone in the area. It is to be anticipated that the system will become nationwide and from any location you will be able to dial any telephone in this country.

A basic plan of such a system is shown in Fig. 2-18. For a large coverage area there is a centrally located group of high-powered transmitters, one for each channel and in any given location, up to a maximum of eight channels. Base receivers spotted conveniently around the area feed signals over wire lines to the central office. The most favorable signal is switched into the telephone system for use. In small-coverage areas it is possible that a single low-powered base transmitter and single co-sited receiver are adequate.

Each mobile unit consists of the transmitter-receiver combination and control unit, plus the channel-switching facilities and signaling circuits. Each unit is equipped to operate on at least all of the channels active in a phone area. For subscribers who are on the road, the mobile set may be equipped with additional channels up to a maximum of eleven. Full duplex operation will be included.

A special and very significant feature of the IMTS system is a feature referred to as multichannel access. A mobile unit automatically seeks an idle channel. Thus, when a mobile unit is activated, it will automatically seek an idle channel and establish a link from mobile to base station. The next call in either direction will then be completed over that channel. When this channel is occupied, all other mobiles in the area will automatically seek another idle base-station channel to establish still another connection. It is anticipated that each

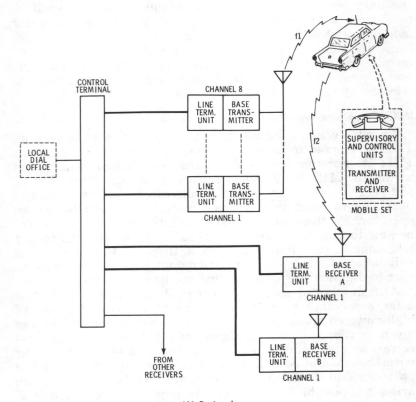

(A) Basic plan.

(B) Subscriber mobile telephone.

Fig. 2-18. IMTS mobile telephone system.

channel will be able to serve a maximum of forty to sixty mobile units. The entire operation will be automatic, and in most cases it will not be necessary to call through a mobile service operator.

LAND TRANSPORTATION RADIO SERVICES

The Land Transportation Radio Services set aside parts of the radio spectrum to be employed for radio communication and control facilities in certain land-transportation operations. Such facilities may not be used either to render a common-carrier communications service or to carry broadcast-program material.

Communications must be related directly to the safety of life or the protection of important property. Permissible communications may also involve the efficient operation of the transportation system described in the application and defined in the rules of eligibility for the particular service.

In general, both the station and operator requirements are the same as those applicable to the Industrial Radio Services.

Station Categories

There are four major categories in the Land Transportation Service; these are motor carrier, railroad, taxicab, and automobile emergency radio services. In the taxicab radio service, certain frequencies are set aside for base and mobile units and for mobile use only. Since there is considerable communication activity in handling a fleet of taxicabs, frequency allocations are often made in pairs, one frequency for the base station and a second frequency for the mobile units. Thus, the base station is able to hear all of the mobile units, but each mobile unit can only receive the base station. A mobile unit does not hear the other mobile units, and thus considerable two-way radio confusion is avoided. Radiocommunication can then be handled more efficiently and effectively from the base station.

Those eligible for licensing in the automobile emergency radio service are associations of owners of private automobiles who provide a private emergency road service for disabled vehicles, and those regularly engaged in the business of providing for the general public, emergency road service for disabled vehicles. Communications must be limited to the safety of life or the protection of important property, and communications required for dispatching repair trucks, tow trucks, or other road-service vehicles to disabled vehicles. Also, associations of owners of private automobiles which provide emergency road service may transmit communications for the purpose of reporting traffic conditions on occasions of abnormal vehicular congestion.

Motor Carrier Radio

An active category of the Land Transportation Service has to do with motor carriers. Frequencies are available for persons engaged in providing a common or contract motor-carrier passenger-transportation service between urban areas, or for those who operate such a service within such an area. Frequencies are available for those persons primarily engaged in providing a common or contract motor-carrier property-transportation service between urban areas or for local distribution or collection of property. Trucking concerns are a principal user of these radio assignments.

One of the earliest motor carrier radio systems is operated by the Lansdale Transportation Company. They operate a 250-watt remote controlled transmitter from their central office. Motor-carrier trucks, repair trucks, and service cars are equipped with mobile stations. The base-station antenna is well over 150 feet above the ground level and reliable communications is maintained within a radius of 30 miles with some consistent communications even beyond this limit.

The control point is shown in Fig. 2-19. Through the window the control-point operator can observe the loading areas below as well as maintain contact with trucks in transit. Microphone and operational controls are in front of the operator. To his left and above are the required monitoring facilities; these include a carrier-operated device and the required aural monitors.

Railroad Radio

Railroads have a variety of applications for two-way radio when regularly engaged in the transportation of passengers or property. Facilities may be used in connection with operation or maintenance including use in connection with motor vehicles engaged in the pickup, delivery, or transfer of property between stations. Since railroads stretch out over considerable distances, there is a need for relay as well as repeater stations.

In lieu of call letters for identification, the railroad radio services may use the name of the railroad and train number, caboose number, engine number, or name of fixed wayside stations. Unit identifiers may also be employed if they are kept on record.

Fig. 2-19. Radio control and traffic-dispatch point of Lansdale Transportation Co.

CITIZENS RADIO SERVICE

The Citizens Radio Service has had a phenomenal growth, as attested by the hundreds of thousands of transmitters now in operation. Segments of the frequency spectrum have been set aside in the Citizens Radio Service to provide for a private short-distance radiocommunications service for the business or personal activities of licensees, for radio signaling, and for the control of remote objects or devices by means of radio. Any person who is eighteen or more years of age (or twelve years for a Class-C station) may obtain a station license in this service if his application meets the requirements of the Citizens Radio Service. Parternerships, associations, trusts, or corporations meeting the requirements of the Citizens Service, including any state, territorial, or local governmental entity, or any service organization or association, including civil defense, may also be licensed.

A station license is required. For Class-A stations, FCC Form 400 must be completed; for station Classes C or D, FCC Form 505 must be completed. You may not operate your Citizens radio equipment under another person's license, nor may you lend your call sign to someone else. No radio operator license is required for the normal operation of a Citizens radio station. However, you are still responsible for knowing and abiding by the laws, rules and regulations, and operating procedures applicable to the Citizens Radio Service.

Citizens radio station users must expect and tolerate interference not only from other legally operating Citizens radio stations, but sometimes from stations legally operating in other radio services. Class-C and -D radio-station users must expect and tolerate interference from industrial, scientific and medical equipment in the 29.96 to 27.28 MHz band. Licensees of Class-A stations must apply for a new authorization before shifting frequency to avoid interference; licensees of Class-C or -D stations may shift to any of the frequencies available in their respective classes without securing any further authorization.

No operator license is required for the operation of a citizens radio station, except that a station manually transmitting Morse code shall be operated by the holder of a third or higher class radiotelegraph operator license. All transmitter adjustments or tests involved in the construction, installation, servicing, or maintenance of a Citizens radio station which may affect the proper operation of such station require a person holding a first- or second-class commercial radio operator license, either radiotelephone or radiotelegraph, as may be appropriate. If energy is radiated during such tests or adjustments, the tests or adjustments must be made by or under the immediate supervision and responsibility of the licensed operator. If the tests or adjustments were made without radiating energy but were made in the absence of a properly licensed person, the transmitter shall be checked for compliance with the technical requirements of the rules by a commercial radio operator of the proper grade before it is placed on the air.

Restrictions on Use

The FCC has had to be very specific in stating those activities that cannot be engaged in by Citizens radio users. There have been numerous improper interpretations of the applicable rules and regulations. Too often regulations have been broken by those who have had just a very short-term, and largely nontechnical association with two-way radio. It is unfortunate that they often flout the law and take it into their own hands instead of using due process in the organization of this new service. The radio spectrum is indeed finite and there are pressing needs for frequency spectra for a great variety of radio services. If the radio spectrum were used in accordance with each individual's personal inclinations, it would indeed reduce radio transmission to a useless hodgepodge. Order and responsibility must prevail if efficient and effective use of the radio spectrum is to be sustained.

Exact statements of the uses that are permitted and those that are prohibited are as follows:

§ 95.81 **Permissible Communications.**

Stations licensed in the Citizens Radio Service are authorized to transmit the following types of communications:

(a) Communications to facilitate the personal or business activities of the licensee.

(b) Communications relating to:

(1) The immediate safety of life or the immediate protection of property in accordance with §95.85.

(2) The rendering of assistance to a motorist, mariner or other traveler.

(3) Civil defense activities in accordance with §95.121.

(4) Other activities only as specifically authorized persuant to §95.87 (regarding operation by other than the licensee).

(c) Communications with stations authorized in other radio services except as prohibited in §95.83.

§ 95.83 **Prohibited Communications.**

(a) A citizens radio station shall not be used:

(1) For any purpose, or in connection with any activity, which is contrary to Federal, State or local law.

(2) For the transmission of communications containing obscene, indecent, or profane words, language, or meaning.

(3) To communicate with an Amateur Radio Service station, an unlicensed station, or foreign stations except for communications pursuant to §95.85(b) and §95.121. [An exception is made in the case of operation of Citizens Radio stations in the United States by Canadians.]

(4) To convey program material for retransmission, live or delayed, on a broadcast facility. Note: A Class A or Class D station may be used in connection with administrative, engineering, or maintenance activities of a broadcasting station;

a Class A or Class C station may be used for control functions by radio which do not involve the transmission of program material; and a Class A or Class D station may be used in the gathering of news items or preparation of programs: Provided, that the actual or recorded transmissions of the Citizens Radio station are not broadcast at any time in whole or in part.

(5) To intentionally interfere with the communications of another station.

(6) For the direct transmission of any material to the public through a public address system or similar means.

(7) For the transmission of music, whistling, sound effects, or any material for amusement or entertainment purposes, or solely to attract attention.

(8) To transmit the word "MAYDAY" or other international distress signals, except when the station is located in a ship, aircraft, or other vehicle which is threatened by grave and imminent danger and requests immediate assistance.

(9) For advertising or soliciting the sale of any goods or services.

(10) For transmitting messages in other than plain language. Abbreviations including nationally or internationally recognized operating signals may be used only if a list of such abbreviations and their meaning is kept in the station records and made available to any Commission representative on demand.

(11) To carry on communications for hire, whether the remuneration or benefit received is direct or indirect.

* * * * * * *

§ 95.91 Duration of Transmissions.

* * * * * * *

(b) All communications between Class D stations (interstation) shall be restricted to not longer than five (5) continuous minutes. At the conclusion of this 5 minute period, or the exchange of less than 5 minutes, the participating stations shall remain silent for at least one minute.

(c) All communication between units of the same Class D station (intrastation) shall be restricted to the minimum practicable transmission.

* * * * * * *

§ 95.95 Station Identification.

*   *   *   *   *   *   *

(c) ... All transmission from each unit of a Citizens Radio station shall be identified by the transmission of its assigned call sign at the beginning and end of each transmission or series of transmissions, but at least at intervals not to exceed ten (10) minutes.

*   *   *   *   *   *   *

It is the licensee's responsibility to see that his equipment is at all times operating in accordance with the rules of the Citizens Radio Service. Off-frequency operation can be guarded against by having measurements made by a properly licensed person who has the frequency-measuring equipment and the skill required to use it. Frequency checks should be made at least each six months. A licensed commercial operator of an appropriate class is required for any adjustments that might affect the proper operation of the system.

The Citizens Radio Service, when properly used, is a valuable communications tool for the professional man (such as the doctor and the engineer), the small business man, and the plain citizen. Improperly used, it can be made useless to everyone because of excessive interference.

Station Categories

The three channel classifications are A, C, and D. Radio control of remote objects or devices comes under the heading of Class-C operation. Class-C operation applies only to radio control, and six specific frequencies between 26.995 and 27.255 MHz are assigned to this service. The maximum power output is 25 watts on 27.255 MHz, and it is limited to 4 watts on the other five channels. Seven additional Class-C channels are available in the 72-76 MHz spectrum.

Frequencies for Class-A stations consist of eight frequencies for base and mobile stations between 462.550 and 462.725 MHz and eight frequencies for mobile stations only between 467.550 and 467.725 MHz. Either amplitude or frequency modulation may be used, and a maximum power output of 50 watts is permissible. Restrictions on antenna structures have to do only with safety to aircraft.

Class-D assignments are the most popular, and they consist of 23 channels located between 26.965 and 27.255 MHz. The class-D frequency assignments (except channels 9 and

11) are given in Table 2-3. The channels in Table 2-3 may be used for communications between units of the same station (intrastation) or different stations (interstation).

The frequency 27.065 MHz (channel 9) is reserved for emergency communications involving the immediate safety of life of individuals or the immediate protection of property. This channel may also be used for communications necessary to render assistance to a motorist. The frequency 27.085 MHz (channel 11) shall be used only as a calling frequency for the sole purpose of establishing communications and moving to another frequency (channel) to conduct communications.

Table 2-3. Class-D CB Channels for Intrastation and Interstation Communications

MHz	Channel
26.965	1
26.975	2
26.985	3
27.005	4
27.015	5
27.025	6
27.035	7
27.055	8
27.075	10
27.105	12
27.115	13
27.125	14
27.135	15
27.155	16
27.165	17
27.175	18
27.185	19
27.205	20
27.215	21
27.225	22
27.255	23

Inasmuch as short-range transmission is intended, the maximum power output for these stations is limited to 4 watts. Furthermore, the antenna of a station at a fixed location must fall within at least one of the following categories:

1. The antenna and its supporting structure does not exceed 20 feet in height above ground level.
2. The antenna and its supporting structure does not exceed by more than 20 feet the height of any natural formation, tree, or manmade structure (not including

a tower or pole) on which the antenna is mounted. Any antenna attached to an existing antenna support structure of another authorized station may exceed neither 60 feet above ground nor the height of the antenna supporting structure of the other station.

3. The antenna is omnidirectional and the highest point of the antenna and its supporting structure does not exceed 60 feet above ground level and the highest point also does not exceed 1 foot in height above the established elevation for each 100 feet of horizontal distance from the nearest point of the nearest airport runway.

Citizens Band Equipment and Uses

The Citizens Radio Service has many uses, and it is particularly attractive because low-cost efficient equipment is

(A) Base unit of family station.

(B) Mobile unit in auto.

Fig. 2-20. Family base-mobile CB installation.

available, and almost everyone is eligible for such a license. The family base-mobile installation is particularly popular. The base station unit (Fig. 2-20A) is installed at home, and a mobile transceiver (Fig. 2-20B) is installed in the family car. When one member of the family is shopping or running errands, he or she is able to maintain contact with home; delays, plan changes, and additional shopping activities can be confirmed and arranged via two-way radio. Often the time of departure from a man's business or place of employment is quite indefinite. At a reasonable distance from home, a mobile-to-home contact can be made to let the famliy know he is on the way home, or possibly to do some last-minute errands on the home trip.

Of course, the two-way radio is handy in case of a car breakdown or other delays. In many localities it is not only possible to call home, but by calling a station operated by a garage or

Courtesy Raytheon Co.

Fig. 2-21. Hand-held CB transceiver.

service station, assistance can be obtained if necessary. This is often quite helpful on long trips, because breakdowns on turnpikes often involve considerable delay if you have to await the routine scheduled trip of a repair truck along the highway.

A calling frequency, channel 11, on the Citizens band has been set up on a nationwide basis. Therefore, a variety of services can be arranged for via two-way radio, such as lodging, repairs, location and traffic guidance, medical assistance, etc. Channel 9 has been set aside for emergency use.

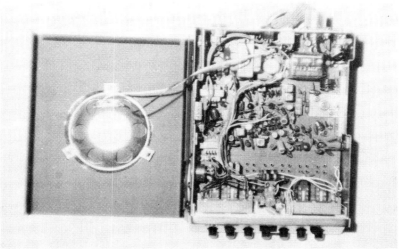

Fig. 2-22. A compact solid-state CB transceiver.

Citizens radio can be a special boon to the small business man. Any type of pickup and/or delivery service can derive benefit in terms of reporting delays, breakdowns, changes in routing and last minute pickups or deliveries. A home oil-delivery service represents only one example of many small businesses that can use Citizens band radio to advantage. Enroute trucks can be informed of call-ins from customers right after the phone calls have been completed and while the trucks may be in the very vicinity of the customer's location. Parts jobbers in the service fields use Citizens radio to advantage in making their deliveries to retail stores and service shops. Communications can even be maintained between the main store and outlying branches of various types of businesses.

Professional people can use Citizens radio for maintaining contact between car and office. Late call-ins or emergencies can be radioed to the doctor as he makes his rounds. A veterinarian making his farm calls can keep in close touch with his office.

There is quite a variety of Citizens radio equipment. Base stations in general are designed to operate from 120-volt ac, while the mobile units operate from the dc of the car battery. Some transceivers can be used with either type of input power. Still other equipment is of the hand-held type (Fig. 2-21), and is battery operated. Conventional batteries can be used in such a unit, or the more recent rechargeable types can be employed. With the latter type it is possible to recharge the batteries overnight using a small charging unit that is driven by 120-volt ac power.

Courtesy Raytheon Co.

Fig. 2-23. CB boat radiotelephone.

Solid-state devices have been a boon to CB equipment design, permitting the design of very compact and lightweight models. The Lafayette Micro-66 model (Fig. 2-22) weighs but 3 lbs-7 oz, and its dimensions are $5\frac{1}{4} \times 7\frac{1}{4} \times 2$ inches. It operates from the car battery, has a 5-watt power input rating, and a receiver sensitivity of 1 microvolt.

Citizens radio can have many public-service functions, being used for liaison work for trade shows, fairs, or other large private or public events. Emergency activities, traffic control, parades, sporting events, etc., can often be expedited with two-way radio gear.

Citizens band operation need not be confined to land transportation. Transceivers can be installed in aircraft or boat (Fig. 2-23). Furthermore, contacts can be made with base stations as well as other types of mobile stations. For example, an aircraft unit can communicate with a unit installed in a car moving along the highway, or a contact can be established between a small pleasure boat and a car driving along the shore.

3

Radio Broadcast Operation

In recent years license holders with grades lower than first-class radiotelephone are being employed as duty operators in various radio broadcast stations. The development of more stable and trouble-free transmitters has made this possible. By passing an appropriate *broadcast endorsement* test, radio announcers and other nontechnical broadcast personnel can perform certain technical duties if the station is properly equipped.

Under the supervision of a first-class radiotelephone license holder you can perform transmitter monitoring duties. One of the attractive facets of the FCC licensing program is that you can become a technical person in radio broadcasting and other radiocommunication fields without involvement in the time and cost of obtaining a baccalaureate degree. You can gain experience, progressing to the second- and then first-class radiotelephone license and, then, develop the skill needed to become a chief engineer. Many radio amateurs have followed this path to a professional career without entering the doors of a school of higher learning.

The endorsement license holder can also serve as a duty operator in standard a-m stations that use directional antennas. Only when the a-m station uses an antenna with a critical directional pattern is the first-class license mandatory. The third-class license holder can also serve as a duty operator in fm broadcast stations, commercial and educational. Operators of a television broadcast station still require a first-class radiotelephone license.

TECHNICAL CONSIDERATIONS

The Federal Communications Commission is concerned with the performance of the transmitter and the technical characteristics of the signal that is radiated from the broadcast antenna. Certain strict technical requirements have been set down with regard to this radio signal so that it may be used effectively by the radio receivers tuned to its frequency. This broadcast signal must not interfere unduly, within the state of the science, with stations operating on other channels. Likewise its radiation must be such that it does not interfere, within certain established interference ratios, with stations in other areas operating on the same frequency. Hence the radiated power output of the broadcast station must be held within FCC specified power limits. If the station uses a directional antenna, that radiation pattern must conform within specified FCC limits.

Each broadcast station must operate on its assigned frequency within a very tight tolerance. Broadcast channels are closely spaced and the broadcast carrier or center frequency must not drift in frequency so as to interfere with stations operating on adjacent channels.

The voice or music signal that is applied to the radio-frequency wave must be added in an efficient and correct manner. Stated technically, the radio-frequency carrier must be properly modulated by a voice or music signal. When a radio-frequency carrier is modulated fully, it is said to be 100% modulated. Only the strong peak-audio passages modulate the carrier to this extent. The average content of voice or music may modulate the radio-frequency carrier by approximately 70%.

If the voice or music components are too strong, in comparison to the strength of the radio-frequency carrier, the transmitter is said to be overmodulated. When the transmitter is overmodulated, the signal, as it is recovered by a radio receiver tuned to the station frequency, may be distorted. When a transmitter is overmodulated, it generates what are called spurious signals. These signals appear on frequencies other than the one which has been assigned to the station. Thus overmodulation of the radio-frequency carrier can cause interference in the reception of broadcast stations using other frequencies.

A transmitter can also be undermodulated. In this case the radio carrier is too strong in comparison to the audio that it is to convey. Under these conditions the broadcast station does not attain its maximum range of transmission and the signals sound weak in all but those locations close to the broadcast antenna.

In summary, the FCC imposes strict requirements in terms of the strength of a signal radiated from the antenna, the frequency of the radiated signal, and whether or not this radio carrier is modulated correctly by the voice and music components.

THE OPERATOR'S RESPONSIBILITIES

The meters and indicators with which the broadcast operator is concerned provide a visual indication of just how well the broadcast transmitter is meeting the technical requirements and considerations. Thus the surveillance of those meters is of significance both in terms of the quality of the broadcast station and the compliance of the station with the FCC technical standards.

The broadcast duty operator must keep a routine watch on operating conditions. He must make appropriate adjustments to keep the meter readings within specified limits. In fact, a regular log of readings must be kept, manually or automatically, to ensure compliance with the FCC rules and regulations. If the readings do not meet or stay within certain limits, appropriate corrections must be made, or in the extreme case, the transmitter must be shut down and the necessary repairs and adjustments made.

Proper attitudes toward technical operating responsibilities are very important, because improper operation may bring an FCC citation or, in the extreme case, result in a fine or loss of license.

TRANSMITTER METERING

There are several key monitoring meters; two of these meters are the final plate voltage and the final plate current. The final plate voltage must be of correct value for operation of the stage that generates the final high-powered radio-frequency carrier. The normal voltage is usually several thousands of volts. If the voltage is too low the final stage will not generate a strong enough rf carrier. If the final plate voltage is too high, power output may be too great or the associated equipment may be damaged.

The final plate-current meter indicates how much current the final rf power amplifier is drawing. It is an indication of how well this stage is operating, and whether or not it is supplying the proper level of power to its output circuit. Most transmitters provide facilities for making adjustments of these quantities so that operation of the transmitter may be set at some point to provide optimum plate voltage and plate current.

A third meter is the power-output meter. It provides a measure of the power that is being transferred from the final stage of the transmitter to the radiating antenna, via the transmission line. This output is read on an rf antenna-current meter, or a calibrated indicator called a reflectometer.

The output-meter reading is very important because it tells just how well the rf power being generated by the transmitter is transferred to the antenna, and how well the antenna system is operating. Depending on system design, the actual reading may change with weather and moisture conditions. Adjustments can usually be made to compensate for weather and terrain effects.

The final meter of importance is the modulation meter. This meter indicates how effectively the voice or music components are modulating the rf carrier. The calibration is given in percentage, showing 100% when the carrier is being fully modulated. A lower level of modulation is indicated by a lower percentage reading. If the voice or music components are made too strong for proper modulation of the rf carrier, the meter will indicate a modulation percentage in excess of 100. Often a flasher or clacking relay will indicate when the transmitter is being overmodulated.

Corrective action must be taken by the operator when there is excessive modulation. Generally, the modulation should not be less than 85% on peaks of frequent recurrence. However, it may be less than 85% when necessary to avoid objectionable loudness. At a-m broadcast stations, modulation must not exceed 100% on negative peaks and 125% on any positive peaks. At fm broadcast stations, modulation must not exceed 100% on either positive or negative peaks.

The antenna ammeter is inserted to measure current into the antenna system. This meter is a direct indication as to whether the station is operating at, above, or below the licensed power. This meter indication should be maintained as close as possible to the licensed operating current, and the operator should know what actions are necessary when the meter indications deviate from that value.

Usually the antenna ammeter is located at the base of the tower and is not easily accessible to the operator on duty.

For this reason, most stations use a remote antenna ammeter which is located at the normal operating position of the person on duty. The remote antenna ammeter indication may be entered in the operating log in lieu of the base current meter indication provided that the remote meter is calibrated once weekly.

Nondirectional a-m stations use a single antenna tower and transmit the radio signal with equal strength in all directions from the station. Directional a-m stations utilize more than one antenna tower. By establishing the position of each tower, the power radiated by each tower, and the phase of the signal in each tower, different signal strengths can be radiated in various directions. Directional antenna systems are used to improve the signal over desired areas and to reduce the signal in the direction of other stations to prevent interference. An operator on duty in a directional a-m station must monitor the tower currents as well as the phase angles among the tower currents. To determine if a directional antenna system is radiating the signal according to a specified radiation pattern, an instrument called an antenna monitor is installed at the station. The antenna monitor enables the operator to determine if the radio-frequency current in each tower is of the correct value and if the phase of the signal radiated by each tower is also correct. Some antenna monitors indicate the ratio of current in each tower to the current in one tower called the reference tower.

If the signal arrives at each tower at the same time, the current in each tower is said to be in phase. In most directional antenna systems, the time the radio-frequency signal reaches each tower from the transmitter is not the same. The rf wave phasing, or time difference, or phase, is measured in degrees. The station license contains a list of the required signal phases and antenna base-current ratios for all the towers in the directional antenna.

The antenna base-current ratio for a tower is calculated by dividing the current-meter reading for that tower by the current-meter reading for the designated reference tower. The ratio, either calculated or read on the antenna monitor, must not deviate more than 5% from the value on the station license (for some antennas, the ratio is extremely critical, and a closer tolerance is stated in the license).

Operators on duty at a-m stations using directional antenna systems should know how to read the antenna-monitor meters. They should also know how to use charts, tables, or other instructions to determine if the station is operating correctly, if attention by the station's first-class operator is required, or if the station must terminate operation.

OPERATING POWER

Each a-m and fm broadcast station is authorized to operate at a specified operating power as indicated on the station license. Operating power for a-m stations is normally the antenna input power, and for fm stations it is the transmitter output power. The operating power must be maintained as near as possible to the value specified by the station license and shall not be more than 105% nor less than 90% of this level. Noncommercial educational fm broadcast stations licensed to operate with transmitter output power of 10 watts or less may be operated at less than the authorized power, but not more than 105% of the authorized power.

Nondirectional a-m broadcast stations employ a single antenna tower. Power determined by the direct method is equal to the product of the antenna resistance and the square of the antenna current:

$$\text{Operating Power} = I_A^2 R_A$$

Directional a-m stations employ multiple radiating elements. Power determined by the direct method for these stations is equal to the product of the resistance common to all antenna towers (called the common-point resistance) and the square of the current common to all antenna towers (called the common-point current).

At fm broadcast stations, if the power is determined by the direct method, it is read directly from the rf transmission-line meter. This meter must be calibrated each six months to give a direct indication of the power supplied to the antenna.

Generally, a-m broadcast stations must determine the operating power by the direct method. Fm broadcast stations may determine the operating power by either method, but most use the indirect method.

For both a-m and fm broadcast stations, operating power determined by the indirect method is equal to the product of the plate voltage and the plate current of the last radio stage, and an efficiency factor:

$$\text{Operating Power} = E_P I_P F$$

When the power of an a-m station is determined by the indirect method, the calculated value is to be entered in the operating log when required transmitter readings are taken.

STATION REQUIREMENTS

When duty operator(s) hold a lesser grade radiotelephone license certain station requirements must be met. Except at times when the station is under the immediate supervision of an operator holding a valid radiotelephone first-class operator license, adjustment of the transmitter equipment by the endorsement license holder shall be limited to the following adjustments:

§ 13.62 Special privileges.

*   *   *   *   *   *   *

(c) The holder of a commercial radiotelegraph first- or second-class license, a radiotelephone second-class license, or a radiotelegraph or radiotelephone third-class permit, endorsed for broadcast station operation may operate any class of standard, fm, or educational fm broadcast station except those using directional antenna systems which are required by the station authorizations to maintain ratios of the currents in the elements of the systems within a tolerance which is less than five percent or relative phases within tolerances which are less than three degrees, under the following conditions:

(1) That adjustments of transmitting equipment by such operators, except when under the immediate supervision of a radiotelephone first-class operator (radiotelephone second-class operator for educational fm stations with transmitter output power of 1000 watts or less), and except as provided in paragraph (d) of this section, shall be limited to the following:

(i) Those necessary to turn the transmitter on and off;

(ii) Those necessary to compensate for voltage fluctuations in the primary power supply;

(iii) Those necessary to maintain modulation levels of the transmitter within prescribed limits;

(iv) Those necessary to effect routine changes in operating power which are required by the station authorization;

(v) Those necessary to change between nondirectional and directional or between differing radiation patterns, provided that such changes require only activation of switches and do not involve the manual tuning of the transmitter's final amplifier or antenna phasor equipment. The switching equipment shall be so arranged that the failure of any relay in the directional antenna system to activate properly will cause the emission of the station to terminate.

(2) The emissions of the station shall be terminated immediately whenever the transmitting system is observed operating beyond the upper and lower limiting values of parameters required to be observed and logged or in any manner inconsistent with the rules or the station authorization, and the above adjustments are ineffective in correcting the condition of improper operation, and a first-class radiotelephone operator is not present.

(3) The special operating authority granted in this section with respect to broadcast stations is subject to the condition that there shall be in employment at the station in accordance with Part 73 of this chapter one or more first-class radiotelephone operators authorized to make or supervise all adjustments, whose primary duty shall be to effect and insure the proper functioning of the transmitting system. In the case of a noncommercial educational fm broadcast station with authorized transmitter output power of 1000 watts or less, a second-class radiotelephone licensed operator may be employed in lieu of a first-class licensed operator.

It is the responsibility of the station licensee to keep transmitter and program logs of station activities. Furthermore, the station licensee must make certain that the person doing the logging and meter reading is properly instructed. When necessary, step-by-step instructions shall be posted for those transmitter adjustments which the lesser grade operator is authorized to make. In the event that the transmitter is observed to be operating in a manner inconsistent with authorization, the transmitter shall be shut down when there is no operator holding a valid first-class radiotelephone license immediately available and the authorized adjustments are not effective in correcting the condition.

LOG REQUIREMENTS

Broadcast stations are required to keep several kinds of logs. The transmitter operator on duty is required to keep the operating (transmitter) log and frequently is responsible for keeping the program log. All logs must be kept by a station employee who is competent to do so having actual knowledge of the facts. Operating and program logs must be signed by the person keeping the logs both at the beginning and end of his period on duty. The logs must be orderly and legible and in such detail that data required is readily available. Each log page must be numbered and dated and times shown in local time. If the area observes "advanced time" during summer months, the log should so indicate.

No log or preprinted log or schedule which becomes a log may be erased or obliterated during the period of required retention. Corrections must be made by striking out the erroneous portion or by a corrective explanation on the log or attached to it. Any corrections or changes, no matter by whom made, must be initialed by the person keeping the log prior to his signing off duty. If changes must be made after the operator has signed off duty, an explanation must be made on the log or an attachment, dated and signed by the person who kept the log or other officials of the station depending on whether the log was a program or operating log.

Program Log

Stations are required to keep program logs of the material broadcast each day. The transmitter operator often must complete the program log in addition to his duties of keeping the transmitter operating log. The program log contains entries identifying each program by name or title and the time the program began and ended. Each program must be identified in the log as to its type and source.

Program log entries for commercial matter must indicate the amount of time devoted to commercial matter during an hourly segment (beginning on the hour) or the duration of each commercial message during the hourly segment. Each sponsored program or announcement must be identified as such during its broadcast, and an indication is required in the log to show this was done. Log entries for each public service announcement must include the name of the organization or interest on whose behalf the announcement was broadcast.

Entries must be made of the time each required station identification was broadcast.

An entry for each announcement of or in behalf of a political candidate is required and must include the name and political affiliation of the candidate.

For fm broadcast stations utilizing a subsidiary communications authorization, there must be a daily log enty to describe the material transmitted on the SCA subchannel. This information may be kept in a special column of the station's regular program log. In the event of a change in the general description of the material transmitted, an entry must be made indicating the time of each such change and a description thereof.

Operating Log

The operator in actual charge of the transmitter at an a-m or fm broadcast station must record certain meter readings in the operating log at the beginning of operation in each mode and at intervals not exceeding three hours. These entries must be the readings observed prior to making any adjustments of the transmitter, and if adjustments are made to restore parameters to their proper operating values, the corrected indications are to be logged. If any parameter deviation is beyond a prescribed tolerance, a notation describing the nature of the corrective action taken to return the parameter to the proper operating value must be entered. Indications of all parameters whose value is affected by modulation of the carrier shall be read without modulation. The operator signs the log when he begins duty and again when going off duty to indicate the time period during which he was in charge of the transmitting equipment. An entry is required for the daily check of the tower lighting system to assure proper illumination of the tower. The time the transmitter begins to supply power to the antenna and the time it ceases are also recorded.

At all standard broadcast (a-m) stations, the following entries shall be made in the operating log:

1. Last-stage plate-voltage meter reading.
2. Last-stage plate-current meter reading.
3. Antenna current or common-point current reading.

At a-m broadcast stations employing directional antenna systems, additional operating-log entries of antenna currents and phases are required.

At fm broadcast stations operating with a transmitter power output above 10 watts, the following entries shall be made in the operating log:

1. Last-stage transmitter plate-voltage meter reading.
2. Last-stage transmitter plate-current meter reading.
3. Rf transmission line meter reading (only fm stations determining the operating power by the direct method need log this).

An fm station operating under a subsidiary communications authorization must keep a log of its operation. The log may be included as part of the operating log for the main channel or may be kept separately. Entries are required of the time the subcarrier generator is turned on and when it is turned off (excluding subcarrier interruptions of 5 minutes or less).

An entry is also required for the time modulation is applied to the subcarrier and the time the modulation is removed from the subcarrier.

Educational fm broadcast stations with an operating power of 10 watts or less are only required to log the time the station begins to supply power to the antenna, the time it stops, and entries concerning the daily observations of tower lights.

STUDY OF RULES AND REGULATIONS

If your interests are in broadcasting you can gain substantial additional knowledge about the field by studying Appendix IV item by item. Do not expect to understand all of the coverage, although each little bit more you grasp will be in your favor when you start your first day of employment in a broadcast station. This added knowledge will also back you up when you go in to take the broadcast-endorsement examination.

STATION PLANS

If a station is to be made acceptable for operation by a lesser grade license holder, be it second-class radiotelephone or third-class endorsement operator, the various indicating meters and operating controls must be made accessible to the operator at his normal on-duty position. Most often his duty involves work as an announcer and/or control-board operator.

A typical station arrangement is shown in Fig 3-1. In this plan the entire station is incorporated into a compact area. The key transmitter readings can be observed when the front panel of the transmitter faces the control console. The equipment rack to the left of the operating position houses the monitoring equipment. From his operating position the control-room operator can also look through a window directly into the studio.

Actually, there are three basic acceptable arrangements. In some installations the transmitter is installed in such a manner that it is visible through an appropriate window (Fig. 3-2). The duty operator can keep a watch on transmitter meters from the operating position at the control console. The modulation monitor is located in an equipment rack to the operator's left.

A second acceptable arrangement (Fig. 3-3) has the control console position looking into the studio. However, the transmitter proper is located in an adjacent room and is not

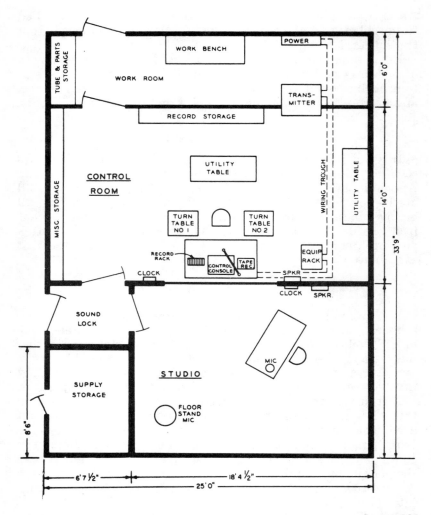

Courtesy RCA

Fig. 3-1. Typical station arrangement.

in the range of vision of the operator on duty. A duplicate set of key meters is mounted in an equipment rack to the rear of the operator (Fig. 3-4). These meters are connected via wires to the metering circuits of the transmitter. The same equipment rack also houses the modulation monitor.

A third acceptable arrangement uses a remote-control facility. In this arrangement the transmitter may be located at a site which is quite distant from the control-console operating position. However, at the operating position there will be a remote-control panel which can be used to control transmitter circuits and operation over interconnecting telephone lines. In this case appropriate metering facilities that can be used to evaluate the transmitter performance from the operating position are installed in the control room. Likewise suitable switches and controls are connected that permit the required transmitter adjustments covered to be made remotely. FCC rules for remote control operation are as follows:

Fig. 3-2. Station WIFI control panel with operator on duty.

§ 73.67 Remote-control operation.

(a) Operation by remote control shall be subject to the following conditions:

(1) The equipment at the operating and transmitting positions shall be so installed and protected that it is not accessible to or capable of operation by persons other than those duly authorized by the licensee.

(2) The control circuits from the operating positions to the transmitter shall provide positive on and off control and shall be such that open circuits, short circuits, grounds or other line faults will not actuate the transmitter and any fault

causing loss of such control will automatically place the transmitter in an inoperative position.

(3) A malfunction of any part of the remote-control equipment and associated line circuits resulting in improper control or inaccurate meter readings shall be cause for the immediate cessation of operation by remote control.

(4) Control and monitoring equipment shall be installed so as to allow the licensed operator at the remote-control point to perform all the functions in a manner required by the Commission's rules.

A monitor for an fm broadcast station is shown in Fig. 3-5. The left meter registers the center-frequency deviation away

Fig. 3-3. Operating position at station WDVR.

from the FCC-assigned carrier frequency. On the fm broadcast frequency this carrier is permitted a maximum drift of 2 kHz. The right meter shows the percentage of fm modulation. To its right is a peak flasher and the associated switch that determines the percentage at which the flasher operates.

A remote-control panel is shown in Fig. 3-6. This is in the studio or control-room unit. A second unit is installed at the transmitter and makes the necessary interconnections that permit the transmitter to be controlled remotely and metered using two telephone lines that link the studio with the transmitter. The three meters permit a measurement of plate voltage, plate current, and antenna current. The transmitter remote-control facility gathers the information from the appropriate transmitter circuits and places a calibrated low voltage on the line to the studio. Proper calibration insures

Fig. 3-4. Meter-monitoring grouping in a control room.

that an accurate reading of the three parameters can be obtained using the three meters of the studio remote-control unit.

The three meters do not record simultaneously but are switched into the circuit by an eleven-position switch. Another switch position gives an indication of tower light circuit operation. Various other positions are available for use in a variety of remote-control, monitoring, and metering applications.

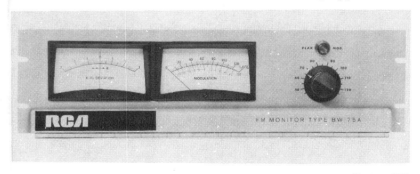

Courtesy RCA

Fig. 3-5. Frequency and modulation monitor.

Courtesy Harris Corp.

Fig. 3-6. Remote metering panel.

Another remote-control panel is shown in Fig. 3-7. In this style, as many as 25 meter readings can be taken over the interconnecting wire lines. To take a sample of a given current at the transmitter, appropriate relays are activated by dialing. This sample will then operate the remote control meter, and its reading can be compared with the normal value shown on the index. The control below the index chart can then be used to make the necessary adjustment to establish a normal reading for the particular circuit under scrutiny.

A directional antenna monitor keeps check on antenna phasing and current levels among the towers. The directional antenna monitor of Fig. 3-8 employs a digital readout. Up

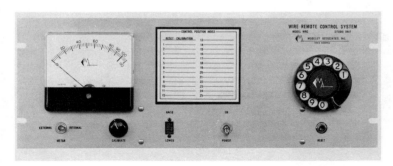

Courtesy Moseley Associates, Inc.

Fig. 3-7. Remote-control panel.

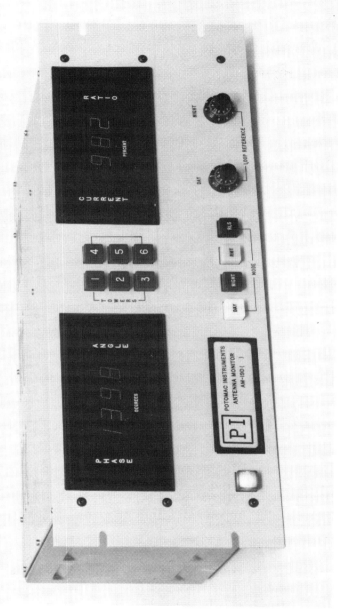

Courtesy Potomac Instruments, Inc.

Fig. 3-8. Antenna monitor.

to six towers can be monitored by depressing the appropriate button. On the left the angle of the antenna current relative to the zero-angle reference is read out. The right readout is that of current ratio of a given tower relative to the current supplied to the reference tower.

FCC RULES AND REGULATIONS

The FCC Study Guide for Element IX suggests the prospective operator be familiar in particular with the following FCC rules and regulations:

73.1	73.111	73.297
73.2	73.112	73.310
73.6	73.113	73.553
73.7	73.116	73.558
73.8	73.119	73.564
73.9	73.253	73.565
73.14	73.258	73.568
73.39	73.263	73.581
73.50	73.264	73.582
73.51	73.265	73.583
73.52	73.267	73.586
73.55	73.268	73.593
73.56	72.274	73.595
73.58	73.281	73.596
73.63	73.282	73.1201
73.67	73.283	73.1206
73.92	73.286	73.1207
73.93	73.289	73.1208
73.95	73.293	73.1211
73.97	73.295	

All of these rules and regulations are given completely in Appendix IV. Study them thoroughly. The following discussions summarize the important points of many of these rules.

Definitions

Standard Broadcast (A-M) Station—A broadcast station licensed for the transmission of radiotelephone emissions primarily intended to be received by the general public, and operated on a frequency in the 535 kHz to 1605 kHz band.

Standard Broadcast Band—The band of frequencies extending from 535 kHz to 1605 kHz.

FM Broadcast Station—A broadcast station transmitting frequency modulated radiotelephone emissions primarily intended to be received by the general public, and operated on a channel in the 88 MHz to 108 MHz band.

FM Broadcast Band—The band of frequencies extending from 88 MHz to 108 MHz.

FM Stereophonic Broadcast—The transmission of a stereophonic program by an fm broadcast station utilizing the main channel and a stereophonic subchannel.

FM Subsidiary Communications Authorization (SCA)—An authorization granted to an fm station for the simultaneous transmission of one or more signals on assigned subcarrier frequencies within the station's assigned channel. Special decoding equipment is required to receive program material furnished on the SCA subchannel. Such material, although broadcast related, is normally intended for reception by paying subscribers.

Daytime—That period of time between local sunrise and local sunset.

Nightime—That period of time between local sunset and local sunrise.

Sunrise and Sunset—For each particular location and during any particular month, the times of sunrise and sunset are specified on most a-m broadcast station licenses. This is necessary because not all standard (a-m) broadcast stations are permitted to operate at night. In order to control objectionable skywave interference, stations which are permitted to operate at night are frequently required to change their modes of operation. These changes may involve the use of directional antenna systems, a reduction in operating power, or both, and normally occur at the sunrise and sunset times listed in the station license.

Broadcast Day—That period of time between local sunrise and 12 midnight local time.

Nominal Power—The power of a standard broadcast station as specified in a system of classification which includes the following values: 50 kW, 25 kW, 10 kW, 5 kW, 2.5 kW, 1 kW, 0.5 kW, and 0.25 kW.

Electrical Terms

Three common electrical terms are volts, amperes, and watts. One thousand volts is a kilovolt (kV), and 1000 watts is a kilowatt (kW). One one-thousandth (1/1000) of an ampere is a milliampere (mA). The operator should know how to use these terms interchangeably. For instance:

2000 volts is the same as 2 kilovolts
0.5 kilovolt is the same as 500 volts
200 watts is the same as 0.2 kilowatt
20 kilowatts is the same as 20,000 watts
0.5 ampere is the same as 500 milliamperes
250 milliamperes is the same as 0.25 ampere

Remote Control Equipment and Operation

Some broadcast stations have the main studio at one location and the transmitter and associated equipment at another location. Rather than have an operator on duty at the transmitter, the controls and metering functions of the transmitter may be at the studio or other location, and the operator on duty may be at this control point. A remote control authorization must be obtained by the licensee from the Commission. Equipment must be installed at the control point that will permit the operator to perform all monitoring and operating functions required by the Commission's rules. If any part of the remote control equipment, meters, or the associated control circuits has a malfunction which results in improper control or meter readings, operation of the transmitter by remote control must cease, and the station can remain on the air only with an operator on duty at the transmitter until the malfunction has been corrected.

Posting of Operator Permits

The operator must post his permit or posting statement at the place where he is on duty at the transmitter control point. FCC Form 759 is a posting statement used by operators employed at more than one station. The permit is posted at one station and FCC Forms 759 are posted at other stations where the operator is employed. The station license and other operating authorizations must be posted at the transmitter control point with all terms visible.

Station Inspection and Availability of Records

The licensee of any radio station shall make the station available for inspection by representatives of the Commission at any reasonable hour. Proofs of performance, logs, measurement records, and other documents required to be maintained must also be available for inspection.

FM Stereo and Subsidiary Communications Authorizations

In addition to monaural operation, fm stations may elect to broadcast stereo programming. This is accomplished by inserting a subchannel within the main fm channel assignment. In addition to the stereophonic subchannel, an entirely different subchannel may also be used. Stations using this subchannel operate under a Subsidiary Communications Authorization (SCA). Programs transmitted by way of the SCA subchannel cannot be received without a special multiplex receiver. Most SCA subcarriers are used for the transmission of subscription background music. Other SCA uses are for detailed weather forecasting, special time signals, and other material of a broadcast nature expressly designed and intended for business, professional, educational, religious, trade, labor, agricultural or other groups engaged in any lawful activity.

Station Identification

Broadcast station-identification announcements shall be made at the beginning and ending of each time of operation and hourly, as close to the hour as feasible, at a natural break in program offerings. Official station identification shall consist of the station's call letters immediately followed by the name of the community or communities specified on the license as the station's location.

Broadcast of Taped or Recorded Material

Any taped or recorded program material in which time is of special significance, or by which an affirmative attempt is made to create the impression that it is occurring simultaneously with the broadcast, must be announced at the beginning as taped or recorded. The language of the announcement shall be clear and in terms commonly understood by the public.

Rebroadcast

The term "rebroadcast" means off-the-air reception by radio of a program originated by another radio station and its simultaneous or subsequent retransmission to the public. No broadcast station shall rebroadcast a program or part of a program of another broadcast station without the permission of the originating station. A copy of the written consent of the licensee originating the program shall be kept by the station rebroadcasting the program.

Sponsor Identification

When a broadcast station transmits matter for which it receives or expects to receive any compensation such as money, services, or other consideration, the station must announce that the matter is sponsored or furnished and also announce the name of the sponsor. However, when a commercial product or service is advertised, use of the sponsor's corporate name, trade name, or name of the product or service will suffice if it is clear that the product name identifies the sponsor. Broadcast station licensees are responsible to insure that their employees or others connected with program material are fully aware that accurate sponsor identification must be obtained and broadcast. Sponsored programs and announcements that do not promote a specific commercial service or product must also be identified by the name of the actual sponsor. Operators keeping program logs must be alert to insure that sponsored matter is announced at the time of the broadcast and the sponsor is correctly identified in the program log.

Lotteries

Broadcast stations are prohibited from transmitting announcements or programs promoting or containing other information that would promote lotteries. A lottery is any scheme in which money or a prize of value is awarded to a person selected by lot or chance, if a condition of winning is that a person must have furnished any money, purchased a particular product or service, or have in his possession a product sold by the sponsor. For example, a door prize given away to a person selected from tickets purchased to gain entry to a particular event is considered a lottery, and the door prize drawing cannot be promoted by radio announcements. Prizes given away to winners selected by lot when "no purchase is necessary" to enter the drawing would not be considered lottery prizes and announcements promoting these "free entry" drawings may be broadcast. Broadcast station operators who are monitoring program material should be aware of the prohibition of broadcasting any type of lottery promotion.

Under certain conditions, advertisements and information regarding state lotteries are exempt from these prohibitions. For details, refer to Section 73.1211 of the Rules and Regulations (Appendix IV).

Broadcast of Telephone Conversations

Before broadcasting a telephone conversation or recording a telephone conversation for later broadcast, the parties to the conversation must be informed of the intention to broadcast the conversation. However, station employees who may be presumed to be aware that their telephone conversations are intended for broadcast are not required to be advised of the broadcasting of their calls. Also, no notice of broadcast is required for persons originating calls to programs which normally broadcast telephone conversations coming into the program.

Emergency Broadcast System and System Test Transmissions

The Emergency Action Notification as it applies to broadcast stations is the notice of the existence of an Emergency Action Condition. An Emergency Action Condition is a national, state, or local area emergency situation posing a threat to the safety of life or property. The FCC provides each broadcast station with a checklist, in summary form, of actions to be taken by the station's operators upon receipt of an Emergency Action Notification, Emergency Action Termination, or Test Messages.

Each station must have in operation at the control point a monitor receiver for receiving the Emergency Action Notification or Termination announcements transmitted by a designated control station. During national level emergency conditions, certain stations will continue operating. Other stations not participating in the national level emergency plan must discontinue operations for the duration of the National Level Emergency Action Condition.

Each station is required to transmit at least once each week an EBS test transmission announcement and signal. All station operators must be thoroughly familiar with the procedures for transmitting the EBS signals and be prepared to take appropriate action in the event of an actual alert. The operator must also be familiar with the operation of the EBS monitor receiver and procedures to follow upon receipt of a test or alert signal from another station.

Tower Lighting

Once each day, the operator is required to check for the proper operation of the tower lighting system. Most tower lighting systems are turned on and off with an automatic actuation switch controlled by a photocell. Stations using the photocell are required to burn the lights continuously if the automatic actuation switch does not operate properly. Any observed or otherwise known outage or improper functioning of a code or rotating beacon light or top light not corrected in 30 minutes should be reported by telephone or telegraph to the nearest Flight Service Station or office of the Federal

Aviation Administration. Notice should also be given to the Flight Service Station or FAA office when the situation has been corrected.

Duplicate and Renewed Commercial Operator Permits

If an operator's license or permit is lost or destroyed, a duplicate license or permit may be obtained by application to the FCC office where the lost or destroyed document was issued. As long as the license or permit has not expired, the operator may continue to operate the station. If the operator's license is required to be posted at his place of duty, a copy of the application, FCC Form 756, submitted for the duplicate license may be posted in lieu of the license pending receipt of the duplicate license.

Operator licenses may be renewed any time during the final year of the license term or within a one year grace period after the expiration of the license by submitting FCC Form 756 to the nearest FCC field office. If the application for renewal is submitted before the license expires, the operator may continue operating unless he hears otherwise from the Commission. A duplicate copy of the renewal application should be posted in lieu of the license, since the license being renewed must accompany the application.

If the renewal application is submitted during the grace period, the license may still be renewed, but the operator has no operating authority between the expiration of the previous license and the issuance of the renewed license.

4

License Procedures And Operator Requirements

The examination for the third-class rediotelephone permit and broadcast endorsement can be taken at a district office of the FCC. A list of offices is given in Table 4-1. Application Form 756 must be completed. This form (Fig. 4-1) plus the appropriate license fee must be submitted to the local office prior to taking the examinaion. An application for a restricted radiotelephone operator permit is made by submitting Form 753 and the $4.00 fee to the FCC office in Gettysburg, Pa. 17325. No examination is required.

The radio spectrum, being finite, must be regulated to provide space for the many services available. Frequencies must be allocated in segments according to the services each sector can best render. Rules of procedure must be established, and assignment and technical regulations enforced, if the universal benefits of communications are to be derived for all.

In the United States the regulatory agency is the Federal Communications Commission (FCC), established under the Communications Act of 1934. Important extracts of this act and the Rules and Regulations appear in the appendices. The United States is also bound by certain international agreements, because radio waves cross national boundaries.

All the extracts in the appendices are of significance in preparing for the commercial radio operator's examinations. More than just a preparation for a test, they emphasize the

Table 4-1. Mailing Addresses for Commission Field Offices

Dist. No.	Office Location	Dist. No.	Office Location
1	1600 Customhouse 165 State St. Boston, MA 02109	12	323-A Customhouse 555 Battery St. San Francisco, CA 94111
2	201 Varick St. New York, NY 10014	13	1782 Federal Office Bldg. 1220 SW 3rd Ave. Portland, OR 97204
3	11425 Federal Courthouse 601 Market St. Philadelphia, PA 19106	14	3256 Federal Bldg. 915 Second Ave. Seattle, WA 98174
4	819 Federal Building 31 Hopkins Plaza Baltimore, MD 21201	15	Suite 2925 The Executive Tower 1405 Curtis St. Denver, CO 80202
5	Military Circle 870 North Military Highway Norfolk, VA 23502	16	691 Federal Building and US Courthouse 316 North Robert Street St. Paul, MN 55101
6	Room 440 1365 Peachtree Street NE Atlanta, GA 30309 **Suboffice:** P.O. Box 8004 Room 238 Federal Bldg. and Courthouse 125 Bull St. Savannah, GA 31402	17	1703 Federal Bldg. 601 East 12th St. Kansas City, MO 64106
		18	Federal Bldg., Room 3935 230 South Dearborn St. Chicago, IL 60604
7	Room 919 51 Southwest First Ave. Miami, FL 33130 **Suboffice:** 738 Federal Office Building 500 Zack St. Tampa, FL 33602	19	1054 Federal Bldg. Washington Blvd. and LaFayette Street Detroit, MI 48226
		20	1305 Federal Bldg. 111 West Huron St. Buffalo, NY 14202
8	829 F. Edward Hebert Federal Bldg. 600 South St. New Orleans, LA 70130	21	502 Federal Bldg. Post Office Box 1021 335 Merchant St. Honolulu, HI 96808
9	New Federal Office Bldg. Room 5636 515 Rusk Ave. Houston, TX 77002 **Suboffice:** 323 Federal Bldg. 300 Willow St. Beaumont, TX 77701	22	Post Office Box 2987 322-323 Federal Bldg. San Juan, PR 00903
		23	Post Office Box 644, Room G63 US Post Office Bldg. Fourth and G St. Anchorage, AK 99510
10	Room 13E7, 1100 Commerce St. Federal Bldg., US Courthouse Dallas, TX 75242	24	Room 411, 1919 M St. NW Washington, DC 20554
11	Suite 501 3711 Long Beach Blvd. Los Angeles, CA 90807 **Suboffice:** Fox Theatre Bldg. 1245 Seventh Ave. San Diego, CA 92101		

responsibilities of a license holder. As a license holder, it is assumed that you know and will abide by the rules and regulations established for the radio service or services with which you are concerned.

Copies of the complete Rules and Regulations for the various radio services are available on subscription from the Superintendent of Documents, U.S. Government Printing Office, Washington, D.C. 20402. If your interest is in two-way radio, you should subscribe to Volume V, which includes aviation, public safety, industrial, and land-transportation radio services. If you have an additional interest in Citizens band radio, Volume VI is available. For the maritime services, there is Volume IV. For broadcast services, Volume III is available. Write the Government Printing Office for prices.

Elements 1 and 2 are about the basic laws and operating practices. The answers given in following chapters to the test questions for elements 1 and 2 are short, concise and yet adequate, to fit in with the direct answers associated with a multiple-choice type of examination. After many answers, the specific laws and regulations given in the appendices are referenced. Read the specific laws carefully. This step will improve your understanding and help you better retain the information you need to know.

Study the book completely, not only the answers to the study-guide questions. Remember that the study-guide questions are only a guide to preparing for the examination. The actual FCC examinations are in the form of multiple-choice questions. These questions are not released and can be changed at will by the Federal Communications Commission. Nevertheless, if you can answer and understand your answers to the study-guide questions, you should have no trouble passing the actual FCC examination.

Notice that safety and distress laws and procedures are stressed. You may never use this information while you are a license holder—but if a distress situation does arise in which you must play a part, there is no more important knowledge that you can have.

The third-class radiotelephone permit is obtained by passing FCC examination elements 1 and 2. Additional element 9 must be passed to obtain the broadcast endorsement. Element 9 relates to basic broadcasting, dealing mostly with definitions, responsibilities, and operating procedures.

Chapters 5, 6, and 7 give the answers to the study-guide questions suggested by the FCC. Know the exact answer to each question. Firm your knowledge by studying each reference to the appendices that follow Chapter 7. In these references, "R.R." means a section in the FCC Rules and Regulations (Appendices III and IV); "SEC." means a section in the Communications Act of 1934, as Amended (Appendix II); and "ART." means an article in the Geneva, 1959, Treaty (Appendix I).

FCC Form 756
October 1975

Form Approved
O.M.B. No. 52-R0080

United States of America
FEDERAL COMMUNICATIONS COMMISSION

APPLICATION FOR RADIO OPERATOR LICENSE
(OTHER THAN AMATEUR LICENSE AND RESTRICTED RADIOTELEPHONE OPERATOR PERMIT)

INSTRUCTIONS	FOR COMMISSION USE ONLY
A. Print or typewrite all information requested and submit one copy to nearest FCC office with one copy of FCC Form 756-B. IF APPLICATION IS FOR DUPLICATE OR REPLACEMENT, SUBMIT TO OFFICE THAT ISSUED ORIGINAL LICENSE. B. Enclose appropriate fee. (DO NOT SEND CASH. Make check or money order payable to Federal Communications Commission.)	

1. I will appear for examination at _____ City _____ State _____
2. Full legal name of applicant (PRINT IN INK OR TYPEWRITE) (First) (Middle) (Last)
3. Permanent Address (No. & Street)
4. Mailing Address (If different from Item 3) (No. & Street)
CITY   STATE   ZIP CODE   CITY   STATE   ZIP CODE

5. Identification

SEX	HEIGHT	WEIGHT	COLOR EYES	COLOR HAIR	DATE OF BIRTH Month Day Year	BIRTH PLACE
Male ☐ Female ☐	....FT. ....IN.					City .... State .... Country ....

CHECK "YES" OR "NO" TO FOLLOWING QUESTIONS AND PROVIDE INFORMATION REQUESTED.   YES NO

6. Are you a Citizen of the United States?

7. Do you have any physical defects such as a speech impediment, blindness, acute deafness, or any other defect which will impair or handicap you in properly using the license for which you are applying? ....
If YES, attach details.

8. Have you been convicted in the last ten years of any crime for which (i) the penalty imposed was a fine of $500 or more, or (ii) for which you were sentenced to imprisonment for more than one year. If the answer is "yes", furnish details for each conviction, giving date, nature of crime, court in which convicted, nature of sentence and where sentence was served. Use separate sheet of paper to provide information.

9. Do you now hold or have you held a commercial operator license, permit or certificate during the last 5 years? ....

10. If the answer to Item 9 is "YES", list license(s) or permit(s) held
(Class)   (Endorsements)   (Serial No. if any)   Date Issued   Place Issued

11. Have you taken within the past two months, a written examination or code test for a commercial operator license or endorsement?
If "YES", give the following information:
Place of Examination _____
Date Examined _____
License or endorsement applied for _____

NOTICE

The Communications Act of 1934 authorizes solicitation of personal information requested in this application. The information is to be used principally to determine if the benefits requested are consistent with the public interest, convenience and necessity.

Commission staff will routinely use the information to evaluate and render a judgement as to whether to grant or deny this application.

If all of the requested information is not provided, the application may be returned without action or processing may be delayed while a request is made to provide the missing information. Therefore, extreme care should be exercised in making certain that all the information requested is provided.

Limited file material may be included in the Commission Computer Facility. Where a possible violation of law is indicated, the records may, as a matter of routine use, be referred to the Commission's General Counsel and forwarded to other appropriate agencies charged with responsibilities of investigating or prosecuting such violations.

THE FOREGOING NOTICE IS REQUIRED BY THE PRIVACY ACT OF 1974, P.L. 93-579, DECEMBER 31, 1974, 5 U.S.C. 552a (e) (3)."

(over)

Fig. 4-1. FCC application form

12. I hereby apply for: *(Check appropriate boxes)*

Class:

☐ New ☐ Radiotelephone ☐ First ☐ Broadcast endorsement
☐ Renewal* ☐ Radiotelegraph ☐ Second ☐ Radar endorsement
☐ Duplicate* ☐ Third ☐ Six months ship radiotelegraph service endorsement
☐ Replacement (See Item 14)* ☐ Aircraft telegraph endorsement

☐ Verification Card (FCC Form 758-F)*

*NOTE: If a current license is held, please attach to application. If it is not attached, explain reason for its absence.

13. If REPLACEMENT license is requested due to name change, complete the following:

	Court Order	Marriage Certificate	Other (attach explanation)
Date issued	....	....	....
Place issued	....	....	....

APPLICANT'S CERTIFICATION

I certify that I am the above-named applicant; that the facts stated in the foregoing application, including the printed declarations therein contained and all exhibits attached thereto, are true of my own knowledge; that I will comply with the instructions pertaining to any operator's examination required in connection with this application; and that I will preserve the secrecy of radio communications as required by law and that I will faithfully adhere to any requirements of law at all times, that this obligation is taken freely, without mental reservation or purpose of evasion, and that I will well and faithfully discharge the duties of the office obtained through my employment under this license if granted.

ANY PERSON WHO WILLFULLY MAKES FALSE STATEMENTS ON THIS FORM CAN BE PUNISHED BY FINE OR IMPRISONMENT. U.S. CODE, TITLE 18, SECTION 1001.

Signature of applicant *Date Signed*

(THE BLANKS BELOW TO BE FILLED IN ONLY BY EXAMINING OFFICER OR ISSUING OFFICE)

EXAMINATION:
☐ NEW _(EXAMINATION DATE)_
☐ RENEWAL
☐ NOT REQUIRED

SPEECH TEST:
☐ SATISFACTORY
☐ UNSATISFACTORY

LICENSE DOCUMENT OR PERMIT ISSUED:
☐ NEW
☐ RENEWED
☐ DUPLICATE
☐ REPLACEMENT
☐ REISSUED DUPLICATE

CODE TEST:
☐ PASSED
☐ FAILED
☐ NOT REQUIRED

TELEGRAPH ENDORSEMENT:
☐ 6 MONTHS MARITIME SERVICE

BROADCAST ENDORSEMENT ☐
AIRCRAFT TELEGRAPH ENDORSEMENT ☐
RADAR ENDORSEMENT ☐

WRITTEN ELEMENT:
ONE ☐ FAILED ☐ PASSED
TWO ☐ FAILED ☐ PASSED
THREE ☐ FAILED ☐ PASSED
FOUR ☐ FAILED ☐ PASSED
FIVE ☐ FAILED ☐ PASSED
SIX ☐ FAILED ☐ PASSED
SEVEN ☐ FAILED ☐ PASSED
EIGHT ☐ FAILED ☐ PASSED
NINE ☐ FAILED ☐ PASSED

758-F ISSUED ☐

☐ NOTIFIED OF FAILURE

REMARKS: _____

(CLASS OF LICENSE)	(NUMBER OF LICENSE)	(DATE OF ISSUE)
(EXAMINER)	(SIGNATURE OF ISSUING OFFICER)	(RADIO DISTRICT)

☆ U.S. GOVERNMENT PRINTING OFFICE: 1975-633-467/158

for radio operator license.

FCC OPERATOR REQUIREMENTS

§ 13.22   Examination requirements.

Applicants for original licenses will be required to pass examinations as follows:

(f) *Radiotelephone third-class operator permit:*
(1) Ability to transmit and receive spoken messages in English.
(2) Written examination elements: 1 and 2.

(h) *Restricted radiotelephone operator permit:*
No oral or written examination is required for this permit. In lieu thereof, applicants will be required to certify in writing to a declaration which states that the applicant has need for the requested permit; can receive and transmit spoken messages in English; can keep at least a rough written log in English or in some other language in general use that can be readily translated into English; is familiar with the provisions of treaties, laws, and rules and regulations governing the authority granted under the requested permit; and understands that it is his responsibility to keep currently familiar with all such provisions.

§ 13.27   Eligibility for reexamination.

An applicant who fails an examination element will be ineligible for 2 months to take an examination for any class of license requiring that element. Examination elements will be graded in the order listed (see § 13.21), and an applicant may, without further application, be issued the class of license for which he qualifies.

§ 13.61   Operating Authority.

(g) *Radiotelephone third-class operator permit.* Any station except:
(1) Stations transmitting television other than Instructional Television Fixed Service stations, or
(2) Stations transmitting telegraphy by any type of the Morse Code, or
(3) Any of the various classes of broadcast stations, or
(4) Class I-B coast stations at which the power is authorized to exceed 250 watts carrier power or 1000 watts peak envelope power, or
(5) Class II-B or Class III-B coast stations, other than those in Alaska, at which the power is authorized to exceed 250 watts carrier power or 1000 watts peak envelope power, or

(6) Ship stations or aircraft stations at which the installation is not used solely for telephony or at which the power is more than 250 watts carrier power or 1000 watts peak envelope power:

Provided, That (1) such operator is prohibited from making any adjustments that may result in improper transmitter operation, and (2) the equipment is so designed that the stability of the frequencies of the transmitter is maintained by the transmitter itself within the limits of tolerance specified by the station license, and none of the operations necessary to be performed during the course of normal rendition of the service of the station may cause off-frequency operation or result in any unauthorized radiation, and (3) any needed adjustments of the transmitter that may affect the proper operation of the station are regularly made by or under the immediate supervision and responsibility of a person holding a first- or second-class commercial radio operator license, either radiotelephone or radiotelegraph as may be appropriate for the class of station involved (as determined by the scope of the authority of the respective licenses as set forth in paragraphs (a), (b), (e), and (f) of this section and § 13.62), who shall be responsible for the proper functioning of the station equipment, and (4) in the case of ship radiotelephone or aircraft radiotelephone stations when the power in the antenna of the unmodulated carrier wave is authorized to exceed 100 watts, any needed adjustments of the transmitter that may affect the proper operation of the station are made only by or under the immediate supervision and responsibility of an operator holding a first- or second-class radiotelephone license, who shall be responsible for the proper functioning of the station equipment.

(h) *Restricted radiotelephone operator permit.* Any station except:

(1) Stations transmitting television, or

(2) Stations transmitting telegraphy by any type of the Morse Code, or

(3) Any of the various classes of broadcast stations other than fm translator and booster stations, or

(4) Ship stations licensed to use telephony at which the power is more than 100 watts carrier power or 400 watts peak envelope power, or

(5) Radio stations provided on board vessels for safety purposes pursuant to statute or treaty, or

(6) Coast stations other than those in Alaska, while employing a frequency below 30 MHz, or

(7) Coast stations at which the power is authorized to exceed 250 watts carrier power or 1000 watts peak envelope power;

(8) At a ship radar station the holder of this class of license may not supervise or be responsible for the performance of any adjustments or tests during or coincident with the installation, servicing or maintenance of the radar equipment while it is radiating energy: *Provided,* That nothing in this subparagraph shall be construed to prevent any person holding such a license from making replacements of fuses or of receiving type tubes:

Provided, That with respect to any station which the holder of this class of license may operate, such operator is prohibited from making any adjustments that may result in improper transmitter operation, and the equipment is so designed that the stability of the frequencies of the transmitter is maintained by the transmitter itself within the limits of tolerance specified by the station license, and none of the operations necessary to be performed during the course of normal rendition of the service of the station may cause off-frequency operation or result in any unauthorized radiation, and any needed adjustments of the transmitter that may affect the proper operation of the station are regularly made by or under the immediate supervision and responsibility of a person holding a first- or second-class commercial radio operator license, either radiotelephone or radiotelegraph, who shall be responsible for the proper functioning of the station equipment.

§ 13.62 Special privileges.

*   *   *   *   *   *   *

(c) The holder of a commercial radiotelegraph first- or second-class license, a radiotelephone second-class license, or a radiotelegraph or radiotelephone third-class permit, endorsed for broadcast station operation may operate any class of standard, fm, or educational fm broadcast station except those using directional antenna systems which are required by the station authorizations to maintain ratios of the currents in the elements of the systems within a tolerance which is less than five percent or relative phases within tolerances which are less than three degrees, under the following conditions:

(1) That adjustments of transmitting equipment by such operators, except when under the immediate supervision of

a radiotelephone first-class operator (radiotelephone second-class operator for educational fm stations with transmitter output power of 1000 watts or less), and except as provided in paragraph (d) of this section, shall be limited to the following:
 (i) Those necessary to turn the transmitter on and off;
 (ii) Those necessary to compensate for voltage fluctuations in the primary power supply;
 (iii) Those necessary to maintain modulation levels of the transmitter within prescribed limits;
 (iv) Those necessary to effect routine changes in operating power which are required by the station authorization;
 (v) Those necessary to change between nondirectional and directional or between differing radiation patterns, provided that such changes require only activation of switches and do not involve the manual tuning of the transmitter's final amplifier or antenna phasor equipment. The switching equipment shall be so arranged that the failure of any relay in the directional antenna system to activate properly will cause the emission of the station to terminate.
 (2) The emissions of the station shall be terminated immediately whenever the transmitting system is observed operating beyond the upper and lower limiting values of parameters required to be observed and logged or in any manner inconsistent with the rules or the station authorization, and the above adjustments are ineffective in correcting the condition of improper operation, and a first-class radiotelephone operator is not present.
 (3) The special operating authority granted in this section with respect to broadcast stations is subject to the condition that there shall be in employment at the station in accordance with Part 73 of this chapter one or more first-class radiotelephone operators authorized to make or supervise all adjustments, whose primary duty shall be to effect and ensure the proper functioning of the transmitting system. In the case of a noncommercial educational fm broadcast station with authorized transmitter output power of 1000 watts or less, a second-class radiotelephone licensed operator may be employed in lieu of a first-class licensed operator.
 (d) When an emergency action condition is declared, a person holding any class of radio operator license or permit who is authorized thereunder to perform limited operation of a standard broadcast station may make any adjustments necessary to effect operation in the emergency broadcast system in accordance with the station's National Defense

Emergency Authorization: *Provided,* That the station's responsible first-class radiotelephone operator(s) shall have previously instructed such person in the adjustments to the transmitter which are necessary to accomplish operation in the Emergency Broadcast System.

§ 1.1102 Payment of fees.

(a) *Filing fees.* Each application or other filing filed on or after August 1, 1970, for which a fee is prescribed in this subpart, must be accompanied by a remittance in the full amount of the filing fee. In no case will an application or other filing be accepted for filing or processed prior to payment of the full amount specified. Filings for which no remittance is received, or for which an insufficient amount is received, shall be returned to the applicant without processing. In the case of multiple applications for which a single check is drawn to cover all fees for the applications, there should be attached to the remittance an accounting sheet or notice stating what fees are covered by the check or money order.

*   *   *   *   *   *   *

§ 1.1117 Schedule of fees for commercial radio operator examinations and licensing.

(a) Except as provided in paragraphs (b) and (c) of this section, applications for commercial radio operator examinations and licensing shall be accompanied by the fees prescribed below:

(1) Applications for new operator license or permit:
    1st-class, 2d-class, or 3d-class, either radiotelephone or radiotelegraph $4
    Provisional radiotelephone 3d-class operator certificate with broadcast endorsement, 1-year term . 2
    Restricted radiotelephone permit 4
    Restricted radiotelephone permit (alien), 5-year term 4
(2) Application for endorsement of license or permit ... 2
(3) Application for renewal of operator license or permit:
    1st-class, 2d-class, or 3d-class, either radiotelephone or radiotelegraph 2
    Restricted radiotelephone operator permit (alien) 4
(4) Application for replacement or duplicate license or permit 2

(5) Application for verification card (form 758-F) 2

(b) Whenever an applicant requests both an operator license or permit and an endorsement, the required fee will be the fee prescribed for the license document involved only.

(c) No fee is required for applications for a replacement license or permit for a marriage-related change of name.

*   *   *   *   *   *   *

5

Element I—Basic Law

I-1. Where and how is an operator license or permit obtained?
—Request an application form and examination schedule from the nearest FCC field office. Submit the application in the prescribed form, including all subsidiary forms and documents accompanied by the prescribed fee, properly completed and signed, in person or by mail, to the office of your choice (usually the closest one). This office will make the final arrangements. (R.R. 13.11a).

I-2. When a licensee qualifies for a higher grade of FCC license or permit, what happens to the lesser grade license?— The license or permit held will be cancelled upon issuance of the new license. (R.R. 13.26).

I-3. Who may apply for an FCC license?—Normally, commercial licenses are issued only to citizens of the United States. (R.R. 13.5a).

I-4. If a license or permit is lost, what action must be taken by the operator?—The commission must be notified immediately and a properly executed application for a duplicate should be submitted. A statement must be included regarding the circumstances involved in the loss of the license. The operator must exhibit in lieu of the original document a signed copy of the submitted application. (R.R. 13.71 and 13.72).

I-5. What is the usual license term for radio operators?— Five years. (R.R. 13.4a).

I-6. What government agency inspects radio stations in the

United States?—The Federal Communications Commission. (Sec. 303n).

I-7. When may a license be renewed?—The application may be filed at any time during the final year of the license term or during a one-year period after the date of expiration of the license. During this one-year grace period any expired license is not valid. (R.R. 13.11a).

I-8. Who keeps the station log?—Person or persons competent to do so having actual knowledge of the facts required. They shall sign the appropriate log when starting duty and again when going off duty. (R.R. 73.111).

I-9. Who corrects errors in the station log?—Only the person originating the entry. (R.R. 73.111, 73.112, and 73.113).

I-10. How may errors in the station log be corrected?—The person originating the entry shall strike out the erroneous portion, initial the correction made and indicate the date of correction. (R.R. 3.111).

I-11. Under what conditions may messages be rebroadcast? —Only by the express authority of the originating station. (Sec. 325a).

I-12. What messages and signals may not be transmitted?— Unnecessary, unidentified, or superfluous communications; obscenity, indecency, or profanity; false signals or any call or signal which has not been assigned by proper authority to the radio station concerned. (R.R. 13.66, 13.67, and 13.68).

I-13. May an operator deliberately interfere with any communication or signal?—No. (R.R. 13.69).

I-14. What type of communication has top priority in the mobile service?—Distress calls, distress messages, and distress traffic. (ART. 37).

I-15. What are the grounds for suspension of operator's license?—Violation of any provision of any act, treaty, or convention binding on the United States; failure to carry out a lawful order of the master or person lawfully in charge of ship or aircraft on which he is employed; damaging or permitting the damage of any radio apparatus or installation; transmission of profane or obscene language, false or deceptive signals, or call signals or letters not assigned to the station in operation; willful or malicious interference with other radiocommunications or frequencies; obtaining, attempting to obtain, or assisting another to obtain or attempt to obtain an operator's license by fraudulent means. (SEC. 303m).

I-16. When may an operator divulge the contents of an intercepted message?—He may divulge the contents of any radiocommunications broadcast or transmission by others for use of the general public or relating to ships in distress. (SEC. 605).

I-17. If a licensee is notified that he has violated an FCC rule or provision of the Communications Act of 1934, what must he do?—He must reply within ten days to the office of the commission originating the official notice. The answer to each notice shall be complete in itself and shall not be abbreviated by reference to other communications or other notices. In every instance the answer shall contain a statement of the action taken to correct the condition or omission complained of, and to preclude its recurrence. Information concerning equipment of concern, and name and license number of operator in charge. (R.R. 1.89).

I-18. If a licensee receives a notice of suspension of his license, what must he do?—No order of suspension of any operator's license shall take effect until 15 days after a notice in writing of the cause has been given to the operator. He may make written application to the commission at any time within said 15 days for a hearing upon such order. If, after a hearing, a license is ordered suspended, the operator shall send his license to the office of the Commission in Washington, D.C., on or before the effective date of such order. (R.R. 1.85).

I-19. What are the penalties provided for violation of the provisions of the Communications Act of 1934 or a rule of the FCC?—Penalties for violation of the Act provide for a fine of not more than $10,000.00 or imprisonment for a term not exceeding one year, or both. The penalty provided for violation of FCC rules is a fine of not more than five hundred dollars ($500) for every day during which said offense occurs. (SEC. 501 and 502).

I-20. Define harmful interference.—Any emission, radiation, or induction which endangers the functioning of a radio-navigation service or other safety services, or seriously degrades, obstructs, or repeatedly interrupts a radiocommunication service operating in accordance with the regulations.

6

Element II—Basic Operating Practice

Each radiotelephone operator should know and abide by the rules and accepted procedures of operation. Courtesy and consideration are two very important factors in minimizing unnecessary interference and maintaining reliable communications on active channels. The radiotelephone license holder should be an example to operators of lower grade, and he should do everything possible to encourage proper operations in the service or services with which he is concerned.

A fine summary of radiotelephone operating practice, recommended by the FCC, follows:

A licensed radio operator should remember that the station he desires to operate should be licensed by the Federal Communications Commission. In order to prevent interference and to give others an opportunity to use the airways, he should avoid unnecessary calls and communications by radio. He should remember that radio signals normally travel outward from the transmitting station in many directions, and therefore these transmitted signals could be intercepted by unauthorized persons.

Before making a radio call, the operator should listen on the communications channel to ensure that interference will not be caused to communications which may already be in progress. At all times in radiocommunications the operator should be courteous.

Station identification should be made clearly and distinctly so as to avoid unnecessary repetition of call letters and enable other stations to clearly identify all calls.

An operator normally exhibits his authority to operate a station by posting a valid operator license or permit at the transmitter control point.

While a radio transmitter is in a public place, it should at all times be either attended by or supervised by a licensed operator; or the transmitter should be made inaccessible to unauthorized persons.

A radio transmitter should not be on the air except when signals are being transmitted. The operator of a radiotelephone station should not press the push-to-talk button, except when he intends to speak into the microphone. Radiation from a transmitter may cause interference even when voice is not transmitted.

When radiocommunications at a station are unreliable or disrupted due to static or fading, it is not a good practice for the operator to continuously call other stations in attempting to make contact, because his calls may cause interference to other stations that are not experiencing static or fading.

A radiotelephone operator should make an effort to train his voice for most effective radiocommunication. His voice should be loud enough to be distinctly heard by the receiving operator, and it should not be too loud, since it may become distorted and difficult to understand at the receiving station. He should articulate his words and avoid speaking in a monotone as much as possible. The working range of the transmitter is affected to some extent by the loudness of the speaker's voice. If the voice is too low, the maximum range of the transmitter cannot be attained; if the voice is too loud, the range may be reduced to zero due to the signals becoming distorted beyond intelligibility. In noisy locations the operator sometimes cups his hands over the microphone to exclude extraneous noise. Normally, the microphone is held from 2 to 6 inches from the operator's lips.

It is important in radiotelephone communications that operators use familiar and well-known words and phrases in order to ensure accuracy and eliminate undue repetition of words. Some radio operating companies, services, networks, associations, etc., select and adopt standard procedure words and phrases for expediting and clarifying radiotelephone conversations. For example in some services, "Roger" means "I have received all of your last transmission." "Wilco" means "Your last message received, understood, and will be complied with," "Out" or "Clear" means "This conversation is ended and no response is expected," "Over" means "My transmission is ended, and I expect a response from you," "Speak slower" means "Speak slowly," "Say Again" means "Repeat," and "Words twice" means "Give every phrase twice."

Often in radiotelephone communications a "phonetic alphabet" or word list is useful in identifying letters or words that may sound like other letters or words of different meanings. For example "group" may sound like "scoop," or "bridge" may sound like "ridge." A phonetic alphabet or word list consists of a list of 26 words each word beginning with a different letter for identifying that particular letter. If the letters in "Group" are represented in a phonetic word by George, Roger, Oboe, Uncle and Peter, the "Group" is transmitted as "Group, G as in George, R as in Roger, O as in Oboe, U as in Uncle, P as in Peter."

In making a call by radio, the call sign or name of the called station is generally given 3 times followed by the call letters of the calling station given 3 times.

In testing a radiotelephone transmitter, the operator should clearly indicate that he is testing, and the station call sign or name of the station, as required by the rules, should be clearly given. Tests should be as brief as possible.

If a radio station is used only for occasional calls, it is a good practice to test the station regularly. Regular tests may reveal defects or faults which, if corrected immediately, may prevent delays when communications are necessary. Caution should be observed by persons testing a station to make certain their test message will not interfere with other communications in progress on the same channel. Technical repairs or adjustments to radiotelephone communication stations are made only by or under the immediate supervision and responsibility of operators holding first- or second-class licenses.

When a licensed operator in charge of a radiotelephone station permits another person to use the microphone and talk over the facilities of the station, he should remember that he continues to bear responsibility for the proper operation of the station.

If an operator wishes to determine the specifications for obstruction marking and lighting of antenna towers he should look in Part 17 of the Rules and Regulations of the FCC. If he wishes to determine the specifications for a particular station, he should examine the station authorization issued by the Commission.

GENERAL (SERIES "O")

II-1. What should an operator do when he leaves a transmitter unattended?—The transmitter should be made inaccessible to unauthorized persons.

II-2. What are the meanings of clear, out, over, roger, words twice, repeat and break?—Refer to the appropriate words in the summary. "Repeat" means to transmit message or section of message again to make certain it has been copied correctly. "Break" means that the transmitting station will pause briefly for an acknowledgment from the station copying his message, or that the receiving station wishes to break in on the transmitting station to request a repeat.

II-3. How should a microphone be treated when used in noisy locations?—The operator can cup his hand over the microphone to exclude extraneous noise.

II-4. What may happen to the received signal when an operator has shouted into a microphone?—The range of the transmission may be decreased, because the signal may become distorted beyond intelligibility.

II-5. Why should radio transmitters be off when signals are not being transmitted?—Radiation from the transmitter may cause interference even though voice or other information is not being transmitted.

II-6. Why should an operator use well-known words and phrases?—It is advisable to ensure accuracy and save time from undue repetition of words.

II-7. Why is the station call sign transmitted?—So that other stations may clearly identify all calls. Call signs must also be given at prescribed times according to the FCC Rules related to the particular radio service.

II-8. Where does an operator find specifications for obstruction marking and lighting where required for the antenna towers of a particular radio station?—Data may be found in Part 17 of the FCC Rules and Regulations.

II-9. What should an operator do if he hears profanity being used at his station?—He should take steps to prevent the profanity from going out on the air.

II-10. When may an operator use a station without regard to certain provisions of the station's license?—In a period of emergency during which normal communications are disrupted. (R.R. 2.405.)

II-11. Who bears the responsibility if an operator permits an unlicensed person to speak over his station?—The licensed operator continues to bear responsibility for the proper operation of the station.

II-12. What is meant by a phonetic alphabet in radiotelephone communications?—It is a list of words helpful in identifying letters or words that may sound like other letters or words that have different meanings.

II-13. How does the licensed operator of a station normally exhibit his authority to operate the station?—A valid operator license must be displayed at the transmitter control point.

II-14. What precautions should be observed when you are testing a station on the air?—Be certain testing does not interfere with other communications in progress on the same channel. Clearly indicate that you are testing. *Technical repairs and adjustments* are the responsibility of the holder of a first- or second-class license.

MARITIME (SERIES "M")

II-M1. What is the importance of the frequency 2182 kHz?—This frequency is the international distress frequency for radiotelephony. It can be used by ships, aircraft, and survival craft using frequencies in the 1605- to 4000-kHz spectrum. This frequency is also the international general radiotelephone calling frequency for the maritime mobile service. It may be used by ship stations and aircraft stations operating in the maritime mobile service. (R.R. 83.352, 83.353a.)

II-M2. Describe completely what action should be taken by a radio operator who hears a distress message; a safety message.—If he is in the vicinity beyond any possible doubt, the operator should immediately acknowledge receipt. However, in areas where reliable communication with one or more coast stations are practicable, a ship station may defer this acknowledgment for a short interval so that a coast station may acknowledge receipt. If not in the vicinity beyond any possible doubt, the operator shall allow a short interval of time to elapse before acknowledging receipt of the message in order to permit stations nearer to the mobile station in distress to acknowledge receipt of the message without interference.

The acknowledgment of the receipt of a distress message is then made according to established radiotelegraph and radiotelephone operating procedures. On order of the master or person responsible for the ship he is on, the operator shall supply the name of his ship and position and the speed at which it is proceeding toward, and the approximate time it will take to reach the station in distress. However, it is important that the station shall be certain that it will not interfere with the emission of other stations better situated to render immediate assistance to the station in distress.

A station may relay the distress message in any of the following cases: (1) station in distress is not able to transmit a distress message, (2) when a responsible person considers that further help is necessary, (3) when the station is not in position to render assistance but has heard that the distress message has not been acknowledged. Read R.R. 83.240, 83.241, and 83.242 carefully.

A safety signal indicates that the station is about to transmit a message concerning the safety of navigation or giving important meteorological warning. Such a station should be given its proper priority of transmission according to ART. 37. (R.R. 83.239, 83.240, 83.241, 83.242, 83.249, and ART. 37.)

II-M3. What information must be contained in distress messages? What procedure should be followed by a radio operator in sending a distress message? What is a good choice of words to be used in sending a distress message?—Distress message should contain following information:

    a. Distress call.
    b. Name of ship.
    c. Geographical position.
    d. Nature of distress.
    e. What kind of assistance is needed.
    f. Any additional information that will help in the rescue, such as type of ship, color, length, etc.

The international radiotelegraph distress signal is S O S (... ---...); the international radiotelephone distress signal is the word "MAYDAY." The radiotelephone distress procedure shall consist of a radiotelephone alarm signal whenever possible, the distress call, and the distress message. This distress transmission shall be made slowly and distinctly, each word being clearly pronounced. Words should be simple and to the point. On advisement, it may be necessary to transmit suitable signals followed by call sign or name to permit direction-finding stations to determine position.

The distress message can be repeated at intervals until an answer is received. If there are no answers, the message may be repeated on any other available frequency on which attention might be attracted.

It is important that you go over each rule and regulation very carefully, word by word. (R.R. 83.234, 83.235, 83.236, and 83.238.)

II-M4. What are the requirements for keeping watch on 2182 kHz? If a radio operator is required to stand watch on an international distress frequency, when may he stop listening?
—Each station on board a ship navigating the Great Lakes and licensed to transmit on telephony within the band 1605 to 3500 kHz shall maintain an efficient watch on 2182 kHz, whenever the station is not being used for transmission on that channel or for communications on other radio channels. Except for stations on board vessels required by law to be fitted with radiotelegraph equipment, each ship station licensed to transmit telephony within the band 1605 to 3500 kHz shall maintain an efficient 2182-kHz watch whenever such station is not being used for transmission on that channel or for communication on other radio channels. (R.R. 83.223.)

II-M5. Under what circumstances may a coast station contact a land station by radio?—To facilitate the transmission or reception of safety communications to or from a ship or aircraft station. (R.R. 81.302a2.)

II-M6. What do distress, safety, and urgency signals indicate? What are the international urgency, safety, and distress signals? In the case of a mobile radio station in distress, what station is responsible for the control of distress-message traffic?—A distress signal indicates that a mobile station is threatened by grave and eminent danger and requests immediate assistance. The urgency signal indicates that the calling station has a very urgent message to transmit concerning the safety of a ship, aircraft, or other vehicle, or the safety of a person. The safety signal indicates that the station is about to transmit a message concerning the safety of navigation or giving important meteorological warning. These signals are as follows:

	Radiotelegraph	Radiotelephone
Distress	S O S	MAYDAY
Urgency	X X X	PAN
Safety	T T T	SECURITY

Whenever possible the mobile station in distress is responsible for the control of distress-message traffic. (R.R. 83.234 through 83.249.)

II-M7. In regions of heavy traffic why should an interval be left between radiotelephone calls? Why should a radio operator listen before transmitting on a shared channel? How long may a radio operator in the mobile service continue attempting to contact a station which does not answer?—An interval should be left between radiotelephone calls to permit the shared use of the channel. An operator should listen before transmitting on a shared channel to make certain that he does not interfere with communications in progress. A call shall not continue for more than 30 seconds. If the called station does not reply, a second call should not be made until after an interval of two minutes. If there is no reply to a call sent three times at intervals of two minutes, the calling shall cease and shall not be renewed until after an interval of fifteen minutes. However, if there is no reason to believe that harmful interference will be caused, the call sent three times at intervals of two minuts may be repeated after a pause of not less than three minutes. (R.R. 83.366 and Summary.)

II-M8. Why are test transmissions sent? How often should they be sent? What is the proper way to send a test message? How often should the station call sign be sent?—If the station is used at infrequent intervals, test transmission can be sent out to ascertain if the transmitter continues to operate in a normal manner. If the station has not been in operation, the transmitter can be checked with a test transmission according to an adopted schedule (once each 24 hours is often considered appropriate). When tests are required, precautions should be taken to prevent any emission that will cause harmful interference. Radiation must be reduced to the lowest practicable value and entirely suppressed if possible. The licensed operator shall listen carefully before test emission so as not to interfere with transmission in progress.

The official call sign of the testing station followed by the word "test" shall be announced as a warning that test emissions are to be made. If any other station transmits the word "wait," testing shall be suspended an appropriate interval of time.

During tests the operator then announces the word "testing" followed by a voice transmission test by numerical count. Test signals should be transmitted that do not conflict with normal operating signals or that will actuate automatic alarms. The test signal shall have a duration not exceeding 10 seconds.

At the conclusion of the test a voice announcement of the official call sign, name of the ship or station, and general location is to be made. It is customary to give the call sign three times. The test transmission shall not be repeated until at least one minute has elapsed or, in a region of heavy traffic, a period of at least five minutes shall elapse before another test transmission is made on key frequencies, (R.R. 83.365 and Summary.)

II-M9. In the mobile service, why should radiotelephone messages be as brief as possible?—Transmissions should be brief to give other stations a chance to use the channel and to minimize interference. (Summary.)

II-M10. What are the meanings of: clear, out, over, roger, words twice, repeat, and break?—Refer to the operating summary at the beginning of this chapter.

II-M11. Does the Geneva 1959 treaty give other countries the authority to inspect US vessels?—Yes, when the mobile station visits another country. The license or a copy certified by the authority which has issued it should be permanently exhibited in the station. (ART 21.)

II-M12. Why are call signs sent? Why should they be sent clearly and distinctly?—Call signs should be sent and transmitted clearly and distinctly so that the station may be identified exactly by other stations and transmission time kept to a minimum. (Summary.)

II-M13. How does the licensed operator of a ship's station exhibit his authority to operate a station?—The original license should be exhibited at a conspicuous place at the principal location on board ship. For certain stations of a portable nature and when the operator holds a restricted radiotelephone operator permit, the person may have on his person either the operator license or a verification card (R.R. 83.165.)

II-M14. When may a code station not charge for messages it is requested to handle?—No charge shall be made for the transmission of distress messages and replies thereto in connection with situations involving the safety of life and property at sea, or for any information concerning danger to navigation as designated. Except for effective tariffs on file with the Commission, no charge shall be made for the service of any public coast station (R.R. 81.179.)

II-M15. What is the difference between calling and working frequencies?—Calling frequencies are used for transmissions from a station solely to secure the attention of another station for a particular purpose. A working frequency is used strictly for carrying on radiocommunications for a purpose other than calling by any station or stations using telegraphy, telephony, of facsimile. (R.R. 83.6.)

ELEMENTS I AND II SELF-TEST

1. When a licensee qualifies for a higher grade license, the lesser grade license
    A. is cancelled.
    B. must be returned to the FCC.
    C. is still valid.
    D. may be used as a verification card.

2. When a license or permit is lost the licensee
    A. must notify FCC within 30 days.
    B. must submit application for a duplicate immediately.
    C. may not operate
    D. is dropped to a lower grade.

3. Radio station inspection is done by the
    A. FBI.
    B. FCC.
    C. FAA.
    D. CIA.

4. Station logs are signed
    A. by FCC inspectors.
    B. when new equipment is installed.
    C. at local sunrise.
    D. when starting and going off duty.

5. An error in the log is corrected by
    A. erasure and dating.
    B. erasure and signing by the person correcting the log.
    C. rewriting the entire page.
    D. striking, dating, and signing by the person correcting the log.

6. Profanity on the air
    A. is legal.
    B. is a violation of FCC rules and regulations.
    C. is permitted because it emphasizes a situation.
    D. must be used on the proper frequency.

7. The urgency signal has
   A. more priority than a distress signal.
   B. more priority than a security signal.
   C. less priority than a safety signal.
   D. a TTT radiotelegraph signal.

8. An operator may divulge the contents of a
   A. distress message.
   B. business transaction.
   C. marine radiotelephone conversation.
   D. commercial shipping instructions.

9. When a licensee receives a notice of suspension it shall take effect
   A. immediately.
   B. in 15 days.
   C. in 30 days.
   D. in 24 hours.

10. A spurious emission that interrupts a radiocommunication service
    A. is permissible at low power.
    B. is permissible on shared channels.
    C. is a strong cw signal.
    D. is harmful interference.

11. "Over" means
    A. change over to a new frequency.
    B. change to another mode of transmission.
    C. change cw speed.
    D. my transmission is ended and a reply is expected.

12. Overmodulation can be caused by
    A. shouting into the microphone.
    B. operating on a wrong frequency.
    C. too little modulator power.
    D. posting a wrong license.

13. Accuracy is improved by
    A. using long descriptive words and sentences.
    B. talking fast.
    C. using well-known words.
    D. shouting into the microphone.

14. Lights of an antenna tower should be inspected every
    A. 24 hours.
    B. 48 hours.
    C. week.
    D. 30 days.

15. A coast station may contact a land station
    A. for a friendly chat.
    B. to handle ship traffic.
    C. to aid in transmission of safety messages.
    D. to relay press traffic.

16. During an emergency
    A. every station should get on the air.
    B. certain provisions of a station license need not be followed when normal communications are disrupted.
    C. shout into the microphone.
    D. chase distress station off your channel.

17. **If in operating on a busy channel you cannot contact the desired station after 7½ minutes of calling**
    A. wait 15 minutes or until propagation conditions improve.
    B. increase power above legal limit.
    C. shout louder into microphone.
    D. change over to sideband.

18. **The phonetic language**
    A. is seldom used in radiocommunications.
    B. helps to clarify phrases.
    C. is a method of clarifying cw.
    D. helps to clarify letters.

19. **When testing**
    A. do it in a hurry and don't waste time listening.
    B. always use maximum assigned power.
    C. be certain testing does not interfere with communications in progress.
    D. shout into the microphone.

20. **When you receive a distress message**
    A. throw your carrier on the air and be ready to help.
    B. try to keep everyone off your channel.
    C. be certain you will not interfere with stations better situated to render assistance.
    D. shut down your station.

21. **An operator required to stand watch on an international distress frequency may stop listening**
    A. when he is handling traffic on another channel.
    B. when it is dark.
    C. when he is 300 miles from shore.
    D. on legal holidays.

22. **The radiotelephone urgency signal is**
    A. MAYDAY.
    B. PAN.
    C. SECURITY.
    D. XXX.

23. **The proper way to start a test message on a clear channel is**
    A. turn on transmitter quickly and shout TEST.
    B. repeat word TESTING and call sign for 20 seconds.
    C. give station call sign and word TEST; then listen on channel before proceeding.
    D. call a friend and tell him to keep channel open.

24. **The word "clear" when used at end of transmission means**
    A. there is to be clear weather.
    B. the channel is free of signals.
    C. propagation is good
    D. communications have been concluded with contacted station.

25. **Call signs**
    A. are a fad of radio operators.
    B. help to identify stations exactly.
    C. can be changed by operator when he gets tired of old one.
    D. are a useless tradition.

26. Who may make application for a radio operator's license?
   A. Citizens.
   B. Nationals of the United States.
   C. FCC certified alien pilots.
   D. All of the above.

27. An operator's license may be renewed
   A. a year before expiration.
   B. within a year after expiration.
   C. within two years after expiration.
   D. as in A and B above.

28. If your channel is active and you have business traffic
   A. put your carrier on the air.
   B. tell others to get off the channel.
   C. you may not deliberately interfere with communications.
   D. call the FCC and say you want a new channel.

29. Which is not ground for license suspension?
   A. Damaging radio apparatus.
   B. Use of indecent language.
   C. Failure to carry out orders of master of ship.
   D. Overmodulation.

30. If a licensee is informed he has violated an FCC rule he must reply within
   A. 24 hours.
   B. a week.
   C. 10 days.
   D. 30 days.

31. What are the penalties for violating the Communications Act?
   A. $500 or one year imprisonment.
   B. $10,000 or one-year imprisonment or both.
   C. $5000 or one-year imprisonment or both.
   D. $500.

32. "Words Twice" means
   A. repeat every word one by one.
   B. give every phrase twice.
   C. use phonetic language.
   D. repeat every letter twice.

33. When operating in a noisy location
   A. shout louder into the microphone.
   B. cup the hands around the microphone.
   C. increase power.
   D. change channel.

34. Where does an operator find specifications for obstruction lights on a radio tower?
   A. Part 17 FCC Rules and Regulations.
   B. FAA office.
   C. Local authority.
   D. Local airport.

35. A call to another station should not continue for more than
   A. 30 seconds.
   B. 20 seconds.
   C. 1 minute.
   D. 2 minutes.

36. An assigned call or identifier must be transmitted
    - A. at the end of each transmission or exchange.
    - B. every 30 minutes.
    - C. only when convenient.
    - D. either A or B.

37. You are called upon to service many stations in the Land-Mobile Services. How can you do the work legally with only one posted license?
    - A. Use any number of photostatic copies.
    - B. Use verification card 758F.
    - C. Use driver's license for identification.
    - D. carry station license.

38. Who may not sign an application for a station license?
    - A. Individual making application.
    - B. Responsible officer for a station to be operated by a company.
    - C. Responsible appointed official for a local government station.
    - D. A friend of applicant who has a first-class license.

39. In periods of emergency
    - A. certain provisions of station license need not be regarded.
    - B. you may not transmit under any circumstances.
    - C. you must call FCC district office.
    - D. reduce transmitter power.

40. All license applications must be
    - A. signed by the applicant.
    - B. notarized.
    - C. sent to National Bureau of Standards.
    - D. sent to FAA

7

Element IX— Broadcast Endorsement

It is essential that the person preparing to take the examination for the broadcast endorsement study in detail and be familiar with the material contained in Chapter 3 and Appendices III and IV. These extracts from the FCC Rules and Regulations include the answers to the following questions.

IX-1. What is meant by the following words or phrases: Standard broadcast station (R.R. 73.1), standard broadcast band (R.R. 73.2), standard broadcast channel (R.R. 73.3), fm station (73.310), fm band (73.310), daytime (R.R. 37.6), nighttime (73.7), broadcast day (73.9) and EBS (73.903)?—Refer to appropriate definition in Appendix IV.

IX-2. Make the following transformation: Kilohertz to hertz, kilovolts to volts, millamperes to amperes—To convert kilohertz to hertz it is necessary to multiply the number of kilohertz by 1000. To convert kilovolts to volts it is necessary to multiply the number of kilovolts by 1000. To convert millamperes to amperes it is necessary to divide the number of milliamperes by 1000.

IX-3. Draw the faces of the following meters and know how to read each: plate current and plate voltage, antenna current and radio-frequency kilowatts, modulation percentage, and antenna phasing in degrees.—Refer to Fig. 7-1. The plate-current meter reading is 270 milliamperes, and the plate

123

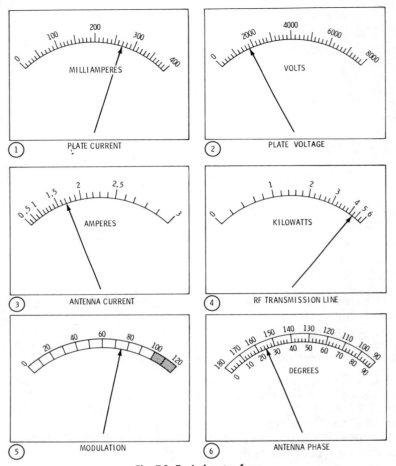

Fig. 7-1. Typical meter faces.

voltage is 1800 volts. Be certain that you agree that this is so by checking the scales of the two meters. The antenna current is 1.7 amperes, and the power in the rf transmission line is 4.5 kilowatts. The modulation percentage is 75%. The phase angle is 25°.

IX-4. What should an operator do if the remote antenna ammeter becomes defective?—A remote antenna ammeter is not required; therefore, authority to operate without it is not necessary if it becomes defective. However, if the remote antenna ammeter does become defective the antenna base current may be read and logged once daily for each mode of operation, pending the return to service of the regular remote meter. (R.R. 73.58.)

IX-5. What should an operator do if the remote control devices at a station so equipped malfunction?—It shall be cause for the immediate cessation of remote-control operation. (R.R. 73.67.)

IX-6. What is the permissible percentage of modulation for a-m and fm stations?—The percentage of modulation shall be maintained as high as possible consistent with good quality of transmission. Generally, it should be no less than 85% on peaks of frequent recurrence, except as necessary to avoid objectionable loudness. For a-m, it shall not exceed 100% on negative peaks of frequent recurrence or 125% on any positive peaks. For fm, it shall not exceed 100% on peaks of frequent recurrence. (R.R. 73.14, 73.55, and 73.268.)

IX-7. What is the permissible frequency tolerance of standard broadcast stations? Of fm stations?—The operating frequency of the a-m broadcast station shall be maintained within 20 Hz of the assigned frequency. The center frequency of each fm broadcast station shall be maintained within 2000 Hz of the center frequency. (R.R. 73.59 and 73.269.)

IX-8. What stations may be operated by a third-class broadcast operator?—Endorsed for broadcast operation, the third-class licensee may operate any standard, fm, or educational fm broadcast station except those using a directional antenna system which are required by the station authorizations to maintain ratios of the currents in the elements of the system within a tolerance which is less than five percent or relative phases within tolerances which are less than three degrees, under prescribed conditions. Study R.R. 13.62 in detail. Also study R.R. 73.93 and R.R. 73.265 carefully.

IX-9. What are the power limitations on broadcast stations?—The operating power of each station will be maintained as near as practicable to the licensed power and shall not exceed the limits of 5% above and 10% below the licensed value, except in an emergency when, due to causes beyond the control of the licensee, it becomes impossible to operate with full power. (R.R. 73.52.)

IX-10. What logs must be kept by broadcast stations according to the Rules and Regulations of the FCC?—Program, operating, and maintenance logs. (R.R. 73.111.)

IX-11. Who keeps the logs?—They must be kept by person or persons competent to do so, having knowledge of the facts required. (R.R. 73.111.)

IX-12. What entries are made in the program log? In the operating log?—Read R.R. 73.112 and 73.113 in Appendix IV.

IX-13. When may abbreviations be used in the station's logs?—Only when proper meaning or explanation is contained elsewhere in the log. (R.R. 13.111.)

IX-14. How and by whom may station logs be corrected?—If a correction is made before the person keeping the log has signed the log upon going off duty, such correction, no matter by whom made, shall be initialed by the person keeping the log prior to his signing of the log when going off duty, as attesting to the fact that the log as corrected is an accurate representation of what was broadcast. If corrections or additions are made on the log after it has been so signed, explanation must be made on the log or on an attachment to it, dated and signed by either the person who kept the log, the appropriate station supervisor, or an officer of the licensee. Any necessary correction shall be made by striking out the erroneous portion or by making a corrective explanation on the log or attachment to it. (R.R. 73.111, 73.112, and 73.113.)

IX-15. According to the Rules and Regulations of the FCC, how long must the station's logs be retained?—For a period of two years. (R.R. 73.115.)

IX-16. What information must be given an FCC inspector at any reasonable hour?—Program, operating, and maintenance logs, equipment performance measurements, a copy of the most recent antenna resistance or common-point impedance measurement submitted to the commission, and a copy of the most recent field intensity measurement to establish performance of the directional antenna. (R.R. 73.116.)

IX-17. What is included in a station identification and how often is it given?—Call letters and location. Station identification shall be made (1) at the beginning and ending of each time of operation and (2) hourly, as close to the hour a feasible, at a natural break in program offerings. Read R.R. 73. 1201 in detail.

IX-18. What should an operator do if the modulation monitor becomes defective?—The station may be operated without the monitor pending its repair or replacement for a period not in excess of 60 days provided appropriate maintenance log entries are made, and the degree of modulation is monitored with an oscilloscope or other acceptable means. (R.R. 73.56.)

IX-19. **When should minor corrections to the transmitter be made, before or after logging the meter reading?**—Logging should be done prior to making any adjustment to the equipment. When appropriate, an indication of the correction made to restore operation to normal should be noted. (R.R. 73.113.)

IX-20. **Should the sponsor's name ever be omitted when reading commercials on the air?**—No. (R.R. 73.119.)

IX-21. **When should an operator announce a program as recorded?**—Whenever the element of time is of significance and the presentation of material would create, either intentionally or otherwise, the impression on the part of the listening audience that the event or program being broadcast is in fact occurring simultaneously with the broadcast. The announcement shall be made at the beginning of the broadcast. (R.R. 73.1208.)

IX-22. **How often should the tower lights be checked for proper operation?**—Once each 24 hours. (R.R. 17.47.)

IX-23. **What record is kept of tower light operation?**—On and off times for each day, time of daily check of proper operation, any observed or otherwise known failure of a tower light, and results of quarterly inspections. (R.R. 17.49.)

IX-24. **What should an operator do if the tower lights fail?**— The failure is to be reported immediately by telephone or telegraph to the nearest Flight Service Station or office of the FAA if not corrected within thirty minutes. (R.R. 17.48.)

IX-25. **What is EBS?**—The Emergency Broadcast System (EBS) is composed of a-m, fm, and tv broadcast stations and nongovernment industry entities operating on a voluntary, organized basis during emergencies at national, state, or operational (local) area levels. (R.R. 73.903.)

IX-26. **What is an emergency situation?**—The condition which exists after the transmission of an emergency action notification and before the transmission of the emergency action termination.

IX-27. **What equipment must be installed in broadcast stations in regard to the reception of an emergency action notification?**—Necessary equipment to receive emergency action notification or termination by specified means. This equipment must be maintained in a state of readiness for reception, including arrangements for human listening watch or automatic alarm devices. (R.R. 73.932.)

IX-28. **How often should EBS test transmissions be sent?— During what time period are they sent?**—They should be made once each week on an unscheduled basis between the hours of 8:30 A.M. and local sunset. (R.R. 73.961 (c).)

IX-29. **During a period of emergency action condition what should all nonparticipating stations do?**—All other broadcast stations will observe broadcast silence. (R.R. 73.919.)

IX-30. **If the tower lights of a station are required to be controlled by a light-sensitive device, and this device malfunctions, when should the tower lights be on?**—They shall burn continuously. (R.R. 17.25 (a) (3).)

The following sample questions are typical of those contained in the element IX examinations. The asterisk indicates the correct answer.

1. Station identification must be given
    - *A. on the hour.
    - B. on the half hour.
    - C. at 30-minute intervals.
    - D. after a commercial.
    - E. after each newscast.

2. The plate voltmeter reads 2 kilovolts. This is the same as
    - A. 0.002 volt.
    - B. 2 volts.
    - C. 20 volts.
    - D. 200 volts.
    - *E. 2000 volts.

3. The plate voltmeter at an fm broadcast station reads 2 kilovolts, and the plate current meter reads 500 milliamperes. With an efficiency of 70 percent, the operating power as determined by the indirect method is
    - A. 140 watts.
    - B. 350 watts.
    - *C. 700 watts.
    - D. 1000 watts.
    - E. 1400 watts.

4. The antenna resistance at a standard broadcast station is 40 ohms, and the antenna current is 5 amperes. The power calculated by the direct method is
    - A. 8 watts.
    - B. 200 watts.
    - C. 900 watts.
    - *D. 1000 watts.
    - E. 8000 watts.

5. Tower one of a three-tower directional antenna array is the reference tower. The antenna current for tower one is 6 amperes. The antenna current for tower two is 2 amperes, and the antenna current for tower three is 3 amperes. The antenna base-current ratio of tower two is
    - *A. 0.33.
    - B. 0.5.
    - C. 1.
    - D. 2.
    - E. e.

It is important that you learn the following terms and definitions.

§ 73.14 Technical definitions.

(a) *Combined audio harmonics.* The term "combined audio harmonics" means the arithmetical sum of the amplitudes of all separate harmonic components. Root sum square harmonic readings may be accepted under conditions prescribed by the Commission.

(b) *Effective field.* The term "effective field" or "effective field intensity" is the root-mean-square (rms) value of the inverse distance fields at a distance of 1 mile from the antenna in all directions in the horizontal plane.

(c) *Nominal power.* "Nominal power" is the power of a standard broadcast station, as specified in a system of classification which includes the following values: 50 kW, 25 kW, 10 kW, 5 kW, 2.5 kW, 1 kW, 0.5 kW, 0.25 kW.

(d) *Operating power.* Depending on the context within which it is employed, the term "operating power" may be synonymous with "nominal power" or "antenna power."

(e) *Maximum rated carrier power.* "Maximum rated carrier power" is the maximum power at which the transmitter can be operated satisfactorily and is determined by the design of the transmitter and the type and number of vacuum tubes used in the last radio stage.

(f) *Plate input power.* "Plate input power" means the product of the direct plate voltage applied to the tubes in the last radio stage and the total direct current flowing to the plates of these tubes, measured without modulation.

(g) *Antenna power.* "Antenna input power" or "antenna power" means the product of the square of the antenna current and the antenna resistance at the point where the current is measured.

(h) *Antenna current.* "Antenna current" means the radio-frequency current in the antenna with no modulation.

(i) *Antenna resistance.* "Antenna resistance" means the total resistance of the transmitting antenna system at the operating frequency and at the point at which the antenna current is measured.

* * * * * * *

(m) Percentage modulation (amplitude):
In a positive direction:

$$M = \frac{MAX - C}{C} \times 100$$

In a negative direction:

$$M = \frac{C - MIN}{C} \times 100$$

where,

M = modulation level in percent,
MAX = instantaneous maximum level of the modulated radio-frequency envelope,
MIN = instantaneous minimum level of the modulated radio-frequency envelope,
C = carrier level of radio-frequency envelope without modulation.

* * * * * * *

§ 73.51 Antenna input power; how determined.

(a) Except in those circumstances described in paragraph (d) of this section, the antenna input power shall be determined by the direct method, i.e., as the product of the antenna resistance at the operating frequency . . . and the square of the unmodulated antenna current at that frequency, measured at the point where the antenna resistance has been determined.

* * * * * * *

(d) The antenna input power shall be determined on a temporary basis by the indirect method . . . in the following circumstances: (1) In an emergency, where the authorized antenna system has been damaged by causes beyond the control of the licensee or permittee . . ., or (2) pending completion of authorized changes in the antenna system, or (3) if changes occur in the antenna system or its environment which affect or appear likely to affect the value of antenna resistance or (4) if the antenna current meter becomes defective. . . . Prior authorization for the indirect determination of antenna input power is not required. However, an appropriate notation shall be made in the operating log.

* * * * * * *

ELEMENT IX SELF-TEST

1. One million hertz is how many kilohertz?
    A. One million.
    B. One thousand.
    C. One hundred.
    D. Ten thousand.

2. The remote antenna meter becomes defective. You must
    A. call FCC.
    B. shut down transmitter.
    C. continue operation.
    D. notify FAA.

3. What deviation constitutes 100% modulation for an fm broadcast transmitter?
   A. ±75 kHz.
   B. ±25 kHz.
   C. 2000 Hz.
   D. 20 Hz.

4. Your modulation meter reads much above 100% on peaks. You must
   A. increase plate voltage.
   B. decrease plate current.
   C. reduce volume level.
   D. readjust modulation meter.

5. A third-class broadcast operator may operate
   A. a directional a-m station with antenna current-ratio tolerance less than 5%.
   B. a radiotelegraph transmitter.
   C. a television station.
   D. none of the above.

6. A broadcast station must keep
   A. only one log.
   B. emergency log only.
   C. no program log.
   D. program and operating logs.

7. An operating log must include
   A. time of sign-on and sign-off.
   B. plate current readings.
   C. service interruptions.
   D. all of above.

8. A program log must not include
   A. name of sponsor.
   B. identification times.
   C. name of announcer.
   D. time of spot commercial.

9. Station logs may be corrected by the
   A. general manager.
   B. chief engineer.
   C. FCC inspector.
   D. person making mistake.

10. FCC inspector must be given upon request
    A. any station log.
    B. income report for the week.
    C. names of all employees and their incomes.
    D. all of above.

11. If a modulation monitor fails
    A. the station must be shut down.
    B. notify the chief engineer.
    C. forget about keeping the log.
    D. turn up the volume level.

12. Scheduled operating logging should be made
    A. each hour only.
    B. after corrective adjustments.
    C. prior to corrective adjustments.
    D. at end of the day only.

13. An announcement of prerecorded material must be made
    A. at sign-off time.
    B. on the half-hour.
    C. to avoid any false impression that program is live.
    D. only on network shows.

14. **What is considered a broadcast day?**
    A. Local sunrise to midnight local time.
    B. Local sunrise to sunset.
    C. 6:00 A.M. to 8:00 P.M.
    D. Daytime.

15. **2000 volts is how many kilovolts?**
    A. 2000.
    B. 2.
    C. 20.
    D. 1/2000.

16. **Frequency of the fm band is**
    A. 540 to 1600 kHz.
    B. 88 to 108 MHz.
    C. 2000 Hz.
    D. ±75 kHz.

17. **If reading on meter 2 (Fig. 7-1) dropped 10%, what would it read?**
    A. 3000.
    B. 1620.
    C. 2000.
    D. 350.

18. **With overmodulation, meter 5 (Fig. 7-1) would read**
    A. 89%.
    B. 100%.
    C. above 100%.
    D. under 85%.

19. **If meter reading of meter 3 (Fig. 7-1) increased 5% it would read**
    A. 1.7.
    B. 1.9.
    C. 2.70.
    D. 1.785.

20. **If the remote control system fails, you should**
    A. shut down the station.
    B. call the chief engineer.
    C. call the FCC inspector.
    D. call the telephone company.

21. **A frequency monitor**
    A. measures carrier drift.
    B. need not be part of remote control monitoring.
    C. is not a compulsory test unit.
    D. All of the above.

22. **Frequency tolerance for an a-m station is**
    A. 20 Hz.
    B. 2000 Hz.
    C. 0.001%.
    D. 0.0001%.

23. **Frequency tolerance for an fm station is**
    A. 0.0001%.
    B. 0.001%.
    C. 2000 Hz.
    D. 20 Hz.

24. **What are the maximum and minimum power limits for a 10-kW a-m station?**
    A. 11 and 9 kW.
    B. 10.5 and 9.5 kW.
    C. 11 and 9.5 kW.
    D. 10.5 and 9 kW.

25. **Who always keeps radio station logs?**
    A. A person competent to do so and with knowledge of facts.
    B. A chief engineer.
    C. A general manager.
    D. An operator at the transmitter.

26. **May abbreviations be used in logs?**
    A. Never.
    B. Anytime.
    C. According to log list.
    D. After approval by FCC inspector.

27. Logs must be retained for
    - A. one year.
    - B. two years.
    - C. five years.
    - D. indefinitely.
28. Call letters and location must be given
    - A. at sign-on.
    - B. at sign-off.
    - C. on the hour.
    - D. A and B above.
29. If frequency meter fails
    - A. station must be shut down.
    - B. notify chief engineer.
    - C. forget about logging it.
    - D. turn up power output.
30. When can sponsor's name be omitted from a commercial?
    - A. Never.
    - B. Upon his request.
    - C. When station manager approves.
    - D. During nonprime time.
31. Tower lights must be checked
    - A. daily.
    - B. weekly.
    - C. monthly.
    - D. hourly.
32. What is the procedure to follow if tower lights fail?
    - A. Shut down station.
    - B. Call FCC immediately.
    - C. Call FAA if repair is not possible in one-half hour.
    - D. Wait for morning.
33. What record must be kept of tower light operation?
    - A. On and off times.
    - B. Failures.
    - C. Daily inspection.
    - D. All of above.
34. What is emergency action condition?
35. During a period of emergency action condition what should nonparticipating stations do?
36. What is percentage modulation indicated by meter A (Fig. 7-2).
    - A. 90.
    - B. −2.
    - C. 100.
    - D. 95.
37. What does meter B (Fig. 7-2) suggest?
    - A. Your fm transmitter is off frequency.
    - B. Everything is just fine with a-m transmitter.
    - C. Problem is arising because a-m transmitter is 20 Hz low.
    - D. Temperature is too low.
38. Arbitrary meter (Fig. 7-2) is reading
    - A. 70.
    - B. 70+.
    - C. high.
    - D. low.
39. Voltmeter (Fig. 7-2) is reading
    - A. 1550 volts.
    - B. too high.
    - C. 1500 volts.
    - D. 1450 volts.
40. Ammeter (Fig. 7-2) is reading
    - A. 325 milliamperes.
    - B. too high.
    - C. 350 milliamperes.
    - D. 350 amperes.

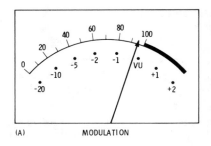

(A) MODULATION

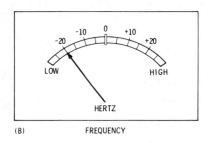

(B) FREQUENCY

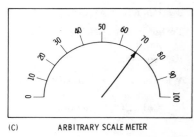

(C) ARBITRARY SCALE METER

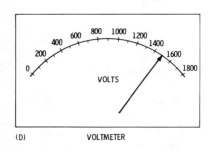

(D) VOLTMETER

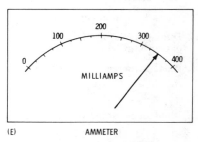

(E) AMMETER

Fig. 7-2. Meters for questions 36 through 40.

41. Indirect power to an antenna is
    A. $I_A^2 R_A$
    B. $I_P^2 E_P$
    C. measured with a field intensity meter.
    D. a substitute method of transmitter power measurement.

42. Direct power measurement is
    A. $I_A^2 R_A$
    B. $I_P E_P$
    C. measured with a field intensity meter.
    D. taken with a plate current meter.

43. The antenna current of a standard broadcast station
    A. is measured in the plate circuit.
    B. is measured with no modulation.
    C. is measured with dummy load.
    D. does not change with modulation.

44. Fm station output power is measured
    A. by indirect means.
    B. by direct means with no output meter.
    C. at the top of the tower.
    D. one each week.

45. Give the definition of a-m modulation percentage.

46. What does antenna resistance mean?

47. Give the definition of antenna power.

48. What is a rebroadcast?

49. Operation by indirect power measurement is permissible when
    A. modulation fails.
    B. the plate circuit current meter fails.
    C. switching to directional antenna.
    D. antenna current meter fails.

50. What is transmitter power input if plate voltage is 3500 volts and dc plate current is 285 milliamperes?
    A. 99.75 watts.
    B. 12.3 kilowatts.
    C. 1 kilowatt.
    D. 9.975 kilowatts.

51. What is the power input on the basis of the readings shown by meters D and E (Fig. 7-2)?
    A. 5250 watts
    B. 428 watts
    C. 525 watts
    D. 4280 watts

52. How many amperes is 300 milliamperes?
    A. 0.3
    B. 3.3
    C. 3
    D. 0.333

53. How many kilowatts is 2000 watts?
    A. 0.2
    B. 2.0
    C. 20
    D. 5

54. The current to tower 2 is 1.5 amperes. What is the current ratio if the current to the reference tower is 2.5 amperes?
    A. 0.166
    B. 0.6
    C. 1.66
    D. 0.375

55. Fm stereo and SCA transmissions
    A. need not be logged.
    B. use directional antennas.
    C. are sent out on subchannels.
    D. require FCC Form 759.

56. Station identification must be made
    A. exactly on the hour.
    B. at end of operations.
    C. after each commercial.
    D. each half hour.

57. To rebroadcast a program of another station you must
   A. obtain a verbal consent.
   B. apply for FCC permission.
   C. notify local FCC office.
   D. retain a written consent.

58. The third-class radio operator at a broadcast station
   A. must be able to tune a crystal oscillator.
   B. conducts visitor tours.
   C. must be able to operate EBS receiver.
   D. keeps the transmitter keys.

59. If you fail the operator examination, you may take the test again after a period of
   A. 90 days.
   B. 30 days.
   C. two months.
   D. one year.

60. Program logs contain
   A. identification of each program.
   B. identification of each commercial.
   C. time devoted to commercials each hour.
   D. all of above.

APPENDIX I

Extracts From the Geneva, 1959, Treaty

SECTION III

TECHNICAL CHARACTERISTICS

PARAGRAPH 93. Harmful interference: Any emission, radiation or induction which endangers the functioning of radionavigation service or of other safety services or seriously degrades, obstructs or repeatedly interrupts a radiocommunication service operating in accordance with these Regulations.

ARTICLE 21
INSPECTION OF MOBILE STATIONS

838 SEC. 1. (1) The governments or appropriate administrations of countries which a mobile station visits, may require the production of the license for examination. The operator of the mobile station, or the person responsible for the station, shall facilitate this examination. The license shall be kept in such a way that it can be produced upon request. As far as possible, the license, or a copy certified by the authority which has issued it, should be permanently exhibited in the station.

ARTICLE 37

PARAGRAPH 1496. The term "communication" as used in this Article means radiotelegrams as well as radiotelephone calls. The order of priority for communications in the mobile service shall be as follows:

1. Distress calls, distress messages, and distress traffic.
2. Communications preceded by the urgency signal.

3. Communications preceded by the safety signal.
4. Communications relating to radio direction-finding.
5. Communications relating to the navigation and safe movement of aircraft.
6. Communications relating to the navigation, movements, and needs of ships, and weather observation messages destined for an official meteorological service.
7. Government radiotelegrams: Priorite' Nations.
8. Government communications for which priority has been requested.
9. Service communications relating to the working of the radio-communications previously exchanged.
10. Government communications other than those shown in 7 and 8 above, and all other communications.

APPENDIX II

Extracts From the Communications Act of 1934, as Amended

SEC. 303 Except as otherwise provided in this Act, the Commission from time to time, as public convenience, interest, or necessity requires shall—(m) (1) Have authority to suspend the license of any operator upon proof sufficient to satisfy the Commission that the licensee

(A) Has violated any provision of any Act, treaty, or convention binding on the United States, which the Commission is authorized to administer, or any regulation made by the Commission under any such Act, treaty, or convention; or
(B) Has failed to carry out a lawful order of the master or person lawfully in charge of the ship or aircraft on which he is employed; or
(C) Has willfully damaged or permitted radio apparatus or installations to be damaged; or
(D) Has transmitted superfluous radio communications or signals or communications containing profane or obscene words, language, or meaning, or has knowingly transmitted
  (1) False or deceptive signals or communications, or
  (2) A call signal or letter which has not been assigned by proper authority to the station he is operating; or
(E) Has willfully or maliciously interfered with any other radio communications or signals; or
(F) Has obtained or attempted to obtain, or has assisted another to obtain or attempt to obtain, an operator's license by fraudulent means.

*   *   *   *   *   *   *

(n) Have authority to inspect all radio installations associated with stations required to be licensed by any Act, or which are subject to the provisions of any Act, treaty, or convention binding on the United

States, to ascertain whether in construction, installations, and operation they conform to the requirements of the rules and regulations of the Commission, the provisions of any Act, the terms of any treaty or convention binding on the United States, and the conditions.

* * * * * * *

SEC. 325 (a) No person within the jurisdiction of the United States shall knowingly utter or transmit, or cause to be uttered or transmitted, any false or fraudulent signal of distress, or communication relating thereto, nor shall any broadcasting station rebroadcast the program or any part thereof of another broadcasting station without the express authority of the originating station.

* * * * * * *

SEC. 501 Any person who willfully and knowingly does or causes or suffers to be done any act, matter, or thing, in this Act prohibited or declared to be unlawful, or who willfully or knowingly omits or fails to do any act, matter, or thing in this Act required to be done, or willfully and knowingly causes or suffers such omission or failure, shall, upon conviction thereof, be punished for such offense, for which no penalty (other than a forfeiture) is provided in this Act, by a fine of not more than $10,000 or by imprisonment for a term not exceeding one year, or both; except that any person, having been once convicted of an offense punishable under this section, who is subsequently convicted of violating any provision of this Act punishable under this section, shall be punished by a fine of not more than $10,000 or by imprisonment for a term not exceeding two years; or both.

SEC. 502 Any person who willfully and knowingly violates any rule, regulation, restriction, or condition made or imposed by the Commission under authority of this Act, or any rule, regulation, restriction, or condition made or imposed by any international radio or wire communications treaty or convention, or regulations annexed thereto, to which the United States is or may hereafter become a party, shall, in addition to any other penalties provided by law, be punished, upon conviction thereof, by a fine of not more than $500 for each and every day during which such offense occurs.

* * * * * * *

SEC. 605 No person receiving or transmitting, or assisting in transmitting, any interstate or foreign communication by wire or radio shall divulge or publish the existence, contents, substance, purport, effect, or meaning thereof, except through authorized channels of transmission or reception, to any person other than the addressee, his agent, or attorney, or to a person employed or authorized to forward such communication to its destination, or to proper accounting or distributing officers of the various communicating centers over which the communication may be passed, or to the master of a ship under whom he is serving, or in response to a subpoena issued by a court of competent jurisdiction, or on demand of other lawful authority; and no person not being authorized by the sender shall intercept any communication and divulge or publish the existence, contents, substance, purport, effect, or meaning of such intercepted communication to any person; and no person not being entitled thereto shall receive or assist in receiving any interstate or foreign communication by wire or radio and use the same or any information therein contained for his own benefit or for the benefit of another not entitled thereto; and no person having received such intercepted communication or having become acquainted with the contents, substance, purport, effect, or meaning of the same or any part thereof, knowing that such information was so obtained, shall divulge or publish the existence, contents, substance, purport, effect, or meaning of the same or any part thereof, or use the same or any information therein contained for his own benefit or for the benefit of another not entitled thereto: *Provided*, That this section shall not apply to the receiving, divulging, publishing, or utilizing the contents of any radio communication broadcast, or transmitted by amateurs or others for the use of the general public, or relating to ships in distress.

APPENDIX III

Extracts From the FCC Rules and Regulations, Parts 1, 2, 13, 81, and 83

This appendix contains reproductions of portions of Parts 1, 2, 13, 81, and 83 of the FCC Rules and Regulations. Even though the information was current at the time this book was printed, the radio operator should have access to and be familiar with a current copy of the Rules and Regulations, since they are subject to revision.

§ 1.85 Suspension of operator licenses.

Whenever grounds exist for suspension of an operator license, as provided in section 303(m) of the Communications Act, the Chief of the Safety and Special Radio Services Bureau, with respect to amateur operator licenses, or the Chief of the Field Engineering Bureau, with respect to commercial operator licenses, may issue an order suspending the operator license. No order of suspension of any operator's license shall take effect until 15 days' notice in writing of the cause for the proposed suspension has been given to the operator licensee, who may make written application to the Commission at any time within said 15 days for a hearing upon such order. The notice to the operator licensee shall not be effective until actually received by him, and from that time he shall have 15 days in which to mail the said application. In the event that physical conditions prevent mailing of the application before the expiration of the 15-day period, the application shall then be mailed as soon as possible thereafter, accompanied by a satisfactory explanation of the delay. Upon receipt by the Commission of such application for hearing, said order of suspension shall be designated for hearing by the Chief, Safety and Special Radio Services Bureau, or the Chief, Field Engineering Bureau, as the case may be, and said order of suspension shall be held in abeyance until the conclusion of the hearing. Upon the conclusion of said hearing, the Commission may affirm, modify, or revoke said order of suspension. If the license is ordered suspended, the operator shall send his operator license to the office of the Commission in Washington, D.C., on or before the effective date of the order, or, if the effective date has passed at the time notice is received, the license shall be sent to the Commission forthwith.

(Sec. 303(m), 48 Stat. 1082, as amended; 47 U.S.C. 303(m))

* * * * * * *

§ 1.89 Notice of violations.

(a) Except in cases of willfulness or those in which public health, interest, or safety requires otherwise, any licensee who appears to have violated any provision of the Communications Act or any provision of this chapter will, before revocation, suspension, or cease and desist proceedings are instituted, be served with a written notice calling these facts to his attention and requesting a statement concerning the matter. FCC Form 793 may be used for this purpose. The Notice of Violation may be combined with a Notice of Apparent Liability to Monetary Forfeiture. In such event, notwithstanding the Notice of Violation, the provisions of § 1.80 apply and not those of § 1.89.

(b) Within 10 days from receipt of notice or such other period as may be specified, the licensee shall send a written answer, in duplicate, direct to the office of the Commission originating the official notice. If an answer cannot be sent nor an acknowledgment made within such 10-day period by reason of illness or other unavoidable circumstances, acknowledgment and answer shall be made at the earliest practicable date with a satisfactory explanation of the delay.

(c) The answer to each notice shall be complete in itself and shall not be abbreviated by reference to other communications or answers to other notices. In every instance the answer shall contain a statement of action taken to correct the condition or omission complained of and to preclude its recurrence. In addition:

(1) If the notice relates to violations that may be due to the physical or electrical characteristics of transmitting apparatus and any new apparatus is to be installed, the answer shall state the date such apparatus was ordered, the name of the manufacturer, and the promised date of delivery. If the installation of such apparatus requires a construction permit, the file number of the application shall be given, or if a file number has not been assigned by the Commission, such identification shall be given as will permit ready identification of the application.

(2) If the notice of violation relates to lack of attention to or improper operation of the transmitter, the name and license number of the operator in charge shall be given.

* * * * * *

§ 2.405 Operation during emergency.

The licensee of any station (except amateur, standard broadcast, FM broadcast, noncommercial educational FM broadcast, or television broadcast) may, during a period of emergency in which normal communication facilities are disrupted as a result of hurricane, flood, earthquake, or similar disaster, utilize such station for emergency communication service in communicating in a manner other than that specified in the instrument of authorization: *Provided:* (a) That as soon as possible after the beginning of such emergency use, notice be sent to the Commission at Washington, D.C., and to the Engineer in Charge of the district in which the station is located, stating the nature of the emergency and the use to which the station is being put, and (b) That the emergency use of the station shall be discontinued as soon as substantially normal communication facilities are again available, and (c) That the Commission at Washington, D.C., and the Engineer in Charge shall be notified immediately when such special use of the station is terminated: *Provided further,* (d) That in no event shall any station engage in emergency transmission on frequencies other than, or with power in excess of, that specified in the instrument of authorization or as otherwise expressly provided by the Commission, or by law: *And provided further,* (e) That any such emergency communication undertaken under this section shall terminate upon order of the Commission.

NOTE : Part 73 of this chapter contains provisions governing emergency operation of standard, FM, noncommercial educational FM, and television broadcast stations. Part 97 of this chapter contains such provisions for amateur stations.

* * * * *

§ 13.4 Term of licenses.

(a) Except as provided in paragraphs (b) and (c) of this section, commercial operator licenses will normally be issued for a term of 5 years from the date of issuance.

(b) Restricted Radiotelephone Operator Permits issued to U.S. citizens or other U.S. nationals will normally be issued for the lifetime of the operator. The terms of all Restricted Radiotelephone Operator Permits issued prior to November 15, 1953, which were outstanding on that date were extended to encompass the lifetime of such operators.

(c) A commercial operator license or permit granted to an alien aircraft pilot under a waiver of the United States nationality provisions of Section 303(1) of the Communications Act will normally be issued for a term of five (5) years from the date of issuance. An operator license or permit issued to an alien aircraft pilot shall be valid only if the operator continues to hold a valid aircraft pilot certificate issued by the Federal Aviation Administration or one of its predecessor agencies.

§ 13.5 Eligibility for new license.

(a) Commercial operator licenses are issued to

(1) United States citizens.

(2) United States nationals.

(3) Citizens of the Trust Territory of the Pacific Islands presenting valid identity certificates issued by the High Commissioner of the Trust Territory.

(4) Aliens holding Federal Aviation Administration pilot certificates.

(5) Aliens holding Federal Communications Commission station licenses, but the operator license will be valid only for operating the station licensed in the name of the alien.

* * * * * *

§ 13.6 Operator license, posting of.

The original license of each station operator shall be posted at the place where he is on duty, except as otherwise provided in this part or in the rules governing the class of station concerned.

* * * * * *

§ 13.7 Operators, place of duty.

(a) Except as may be provided in the rules governing a particular class of station, one or more licensed radio operators of the grade specified by this part shall be on duty at the place where the transmitting apparatus of each licensed radio station is located and in actual charge thereof whenever it is being operated: *Provided, however,* That (1) subject to the provisions of paragraph (b) of this section, where remote control of the transmitting apparatus has been authorized to be used, the Commission may modify the foregoing requirements upon proper application and showing being made so that such operator or operators may be on duty at the control point in lieu of the place where the transmitting apparatus is located; (2) in the case of two or more stations, except broadcast, licensed in the name of the same person to use frequencies above 30 megahertz only, a licensed radio operator holding a valid radiotelegraph or radiotelephone first- or second-class license who has the station within his effective control may be on duty at any point within the communication range of such stations in lieu of the transmitter location or control point during the actual operation of the transmitting apparatus and shall supervise the emissions of all such stations so as to insure the proper operation in accordance with the station license.

(b) An operator may be on duty at a remote control point in lieu of the location of the transmitting apparatus in accordance with the provisions of paragraph (a) (1) of this section: *Provided*, That all of the following conditions are met: (1) The transmitter shall be so installed and protected that it is not accessible to other than duly authorized persons; (2) the emissions of the transmitter shall be continuously monitored at the control point by a licensed operator of the grade specified for the class of station involved; (3) provision shall be made so that the transmitter can quickly and without delay be placed in an inoperative condition by the operator at the control point in the event there is a deviation from the terms of the station license; (4) the radiation of the transmitter shall be suspended immediately when there is a deviation from the terms of the station license.

§ 13.8 **Provisional Radio Operator Certificate.**

(a) In circumstances requiring immediate authority to operate a radio station pending submission of proof of eligibility or of qualifications or pending a determination by the Commission as to these matters, an applicant for a radio operator license may be issued a Provisional Radio Operator Certificate.

(b) Except as provided by paragraph (d) of this section, if the Commission finds that the public interest will be served, it may issue such certificates for a period not to exceed 6 months with such additional limitations as may be indicated.

(c) Except as provided by paragraph (d) of this section, a Provisional Radio Operator Certificate will not be issued if the applicant has not fulfilled examination or service requirements, if any, for the license applied for.

(d) A request for a Provisional Radio Operator Certificate for a radiotelephone third-class operator permit endorsed for broadcast station operation shall be made on FCC Form 756C, which provides for a certification by the holder of a radiotelephone first-class operator license that he is responsible for the technical maintenance of a radio broadcast station, and that he has instructed the applicant in the operation of a broadcast station and believes him to be capable of performing the duties expected of a person holding a radiotelephone third-class operator permit with broadcast station operation endorsement. If the Commission finds that the public interest will be served, it may issue such certificates under the following conditions:

(1) The certificate may be issued for a period not to exceed 12 months.

(2) The certificate is not renewable.

(3) The certificate may be issued to a person only once.

(4) Additional limitations may be specified, as necessary.

(5) The certificate may be issued prior to the fulfillment of examination requirements for the radiotelephone third-class operator permit endorsed for broadcast station operation.

APPLICATIONS

§ 13.11 **Procedure.**

(a) *General.* Applications shall be governed by applicable rules in force on the date when application is filed (see § 13.28). The application in the prescribed form and including all required subsidiary forms and documents, properly completed and signed, and accompanied by the fee prescribed in the Commission's general fee schedule as set forth in Subpart G, Part 1 of this chapter, shall be submitted to the appropriate office as indicated in paragraph (b) of this section. If the application is for renewal of license, it may be filed at any time during the final year of the license term or during a 1-year period of grace after the date of expiration of the license sought to be renewed. During this 1-year period of grace, an expired license is not valid. A renewed license issued upon the basis of an application filed during the grace period will be dated currently and will not be backdated to the date of expiration of the license being renewed. A renewal application shall be accompanied by the license sought to be renewed. If the prescribed service requirements for renewal without examination (see § 13.28) are fulfilled, the renewed license may be issued by mail. If the service record on the reverse side of the license does not fully describe or cover the service desired by the applicant to be considered in connection with license renewal (as might occur in the case of service rendered at U.S. Government stations), the renewal application shall be supported by documentary evidence describing in detail the service performed and showing that the applicant actually performed such service in a satisfactory manner. A separate application must be submitted for each license involved, whether it requests renewal, new license, endorsement, duplicate, or replacement.

(b) *Place of filing.* (1) Applications for Restricted Radiotelephone Operator Permits shall be filed as follows:

(i) United States citizens, United States nationals, citizens of the Trust Territory of the Pacific Islands— File application FCC Form 753:

(*a*) When there is an immediate need for the permit for safety purposes, submit the application in person or by an agent to the nearest field office of the Field Operations Bureau—Federal Communications Commission. The application must be accompanied by a written showing by the applicant that he has an immediate need for the permit for safety purposes.

(*b*) When the applicant is located in Alaska, Hawaii, Puerto Rico, or the Virgin Islands of the United States, submit the application in person or by mail to the nearest field office of the Field Operations Bureau— Federal Communications Commission.

(*c*) All other cases—mail the application to the Federal Communications Commission, Gettysburg, Pennsylvania 17325.

(ii) Aliens holding a station license—File application FCC Form 755:

(*a*) When there is an immediate need for the permit for safety purposes, submit the application in person or by an agent to the nearest field office of the Field Operations Bureau—Federal Communications Commission. The application must be accompanied by a written showing by the applicant that he has an immediate need for the permit for safety purposes.

(*b*) When the applicant is located in Alaska, Hawaii, Puerto Rico, or the Virgin Islands of the United States, submit the application in person or by mail to the nearest field office of the Field Operations Bureau—Federal Communications Commission.

(c) All other cases—mail application to the Federal Communications Commission, Washington, D.C. 20554.

(iii) Alien aircraft pilots holding Federal Aviation Administration pilot certificates—Submit the application FCC Form 755 ether in person or by mail to the Federal Communications Commission, Washington, D.C. 20554 (see § 13.4(c)).

(2) An application for an operator license or permit of any other class, or for a verification card, shall be submitted in person or by mail to the field office at which the applicant desires his application to be considered and acted upon, and which office will make final arrangements for conducting any required examination. Whenever an examination is required to be taken at a designated examination point away from a field office, the application shall be submitted in advance of the examination to the field office having jurisdiction over the area in which the examination is to be given.

(3) The form entitled "Verification of Operator License or Permit" (FCC 759) may be obtained from any of the Commission's field offices. The certification under Part B of the form shall be completed by the licensee or general manager of the radio station where the statement is to be posted. When the FCC Form 759 is properly validated, it may be posted in lieu of the original radio operator license or permit when the holder of that license or permit is employed at more than one station.

(c) *Restricted radiotelephone operator permit.* No oral or written examination is required for this permit. If the application is properly completed and signed, and if the applicant is found to be qualified, the permit may be issued forthwith.

(d) *Short term license.* A license or permit issued for a term of less than five years (see § 13.4), may be renewed without further examination, provided proper application is filed in accordance with paragraph (a) of this section.

(e) Blind applicant. A blind person seeking an examination for radiotelephone first-class operator license, radiotelephone second-class operator license, radiotelephone third-class operator license, and radiotelephone third-class operator license with broadcast station operation endorsement shall make a request in writing to the appropriate field office for a time and date to appear for such examination. The examination shall be administered only at the field office. Requests for examinations shall be made at least 2 weeks prior to the date on which the examination is desired.

§ 13.12 Special provisions, radiotelegraph first class.

An applicant for a radiotelegraph first-class operator license must be at least 21 years of age at the time the license is issued and shall have had an aggregate of one year of satisfactory service as an operator manipulating the key of a manually operated public ship or coast station handling public correspondence by radiotelegraphy.

§ 13.13 Age limit, restricted radiotelephone operator permit.

An applicant for a restricted radiotelephone operator permit must be at least 14 years of age at the time the permit is issued.

EXAMINATIONS

§ 13.21 Examination elements.

(a) Written examinations will comprise questions from one or more of the following examination elements:

 1. *Basic law.* Provisions of laws, treaties and regulations with which every operator should be familiar.

 2. *Basic operating practice.* Radio operating procedure and practices generally followed or required in communicating by means of radiotelephone stations.

 3. *Basic radiotelephone.* Technical, legal and other matters applicable to the operation of radiotelephone stations other than broadcast.

 4. *Advanced radiotelephone.* Advanced technical, legal and other matters particularly applicable to the operation of the various classes of broadcast stations.

 5. *Radiotelegraph operating practice.* Radio operating procedures and practices generally followed or required in communicating by means of radiotelegraph stations primarily other than in the maritime mobile services of public correspondence.

 6. *Advanced radiotelegraph.* Technical, legal and other matters applicable to the operation of all classes of radiotelegraph stations, including operating procedures and practices in the maritime mobile services of public correspondence, and associated matters such as radio navigational aids, message traffic routing and accounting, etc.

 7. *Aircraft radiotelegraph.* Basic theory and practice in the operation of radio communication and radio navigational systems in general use on aircraft.

 8. *Ship radar techniques.* Specialized theory and practice applicable to the proper installation, servicing and maintenance of ship radar equipment in general use for marine navigational purposes.

 9. *Basic broadcast.* Basic regulatory and elementary technical matters applicable to the operation of standard, commercial FM, and noncommercial educational FM broadcast stations.

* * * * * * *

§ 13.22 Examination requirements.

Applicants for original licenses will be required to pass examinations as follows:

(a) *Radiotelephone second-class operator license:*
(1) Ability to transmit and receive spoken messages in English.
(2) Written examination elements: 1, 2, and 3.

(b) *Radiotelephone first-class operator license:*
(1) Ability to transmit and receive spoken messages in English.
(2) Written examination elements: 1, 2, 3, and 4.

(c) *Radiotelegraph second-class operator license:*
(1) Ability to transmit and receive spoken messages in English.
(2) Transmitting and receiving code test of twenty (20) words per minute plain language and sixteen (16) code groups per minute.
(3) Written examination elements: 1, 2, 5, and 6.

(d) [Reserved]

(e) *Radiotelegraph first-class operator license:*

(1) Ability to transmit and receive spoken messages in English.

(2) Transmitting and receiving code test of twenty-five (25) words per minute plain language and twenty (20) code groups per minute.

(3) Written examination elements: 1, 2, 5, and 6.

(f) *Radiotelephone third-class operator permit:*

(1) Ability to transmit and receive spoken messages in English.

(2) Written examination elements: 1 and 2.

(g) *Radiotelegraph third-class operator permit:*

(1) Ability to transmit and receive spoken messages in English.

(2) Transmitting and receiving code test of twenty (20) words per minute plain language and sixteen (16) code groups per minute.

(3) Written examination elements: 1, 2, and 5.

(h) *Restricted radiotelephone operator permit:*

No oral or written examination is required for this permit. In lieu thereof, applicants will be required to certify in writing to a declaration which states that the applicant has need for the requested permit; can receive and transmit spoken messages in English; can keep at least a rough written log in English or in some other language in general use that can be readily translated into English; is familiar with the provisions of treaties, laws, and rules and regulations governing the authority granted under the requested permit; and understands that it is his responsibility to keep currently familiar with all such provisions.

§ 13.23 Examination form.

The written examination shall be in English, except when waived in accordance with authority specified in section 0.314. In the case of a blind applicant, the examination questions shall be read orally and the dictated answers recorded by a Commission examiner authorized to administer such oral examination.

§ 13.24 Passing mark.

A passing mark of 75 percent of a possible 100 percent will be required on each element of a written examination.

§ 13.25 New class, additional requirements.

The holder of a license, who applies for another class of license, will be required to pass only the added examination requirements for the new class of license: *Provided,* That the holder of a radiotelegraph third-class operator permit who takes an examination for a radiotelegraph second-class operator license more than one year after the issuance date of the third-class permit will also be required to pass the code test prescribed therefor: *Provided further,* That no person holding a new, duplicate, or replacement restricted radiotelephone operator permit issued on the basis of a declaration, or a renewed restricted radiotelephone operator permit which renews a permit issued upon the basis of a declaration shall, by reason of the declaration or the holding of such permit, be relieved in any respect of qualifying by examination when applying for any other class of license.

§ 13.26 Canceling and issuing new licenses.

If the holder of a license qualifies for a higher class in the same group, the license held will be canceled upon the issuance of the new license. Similarly, if the holder of a restricted operator permit qualifies for any other class or type of license or permit, the Restricted Radiotelephone Operator Permit will be cancelled upon issuance of the new authorization, except as provided in section 13.3(b).

§ 13.27 Eligibility for re-examination.

An applicant who fails an examination element, including a code test element, will be ineligible for 2 months to take an examination for any class of license requiring that element. Examination elements will be graded in the order listed (see § 13.21), and an applicant may, without further application, be issued the class of license for which he qualifies.

NOTE: A month after date is the same day of the following month, or if there is no such day, the last day of such month. This principle applies for other periods. For example, in the case of the 2-month period to which this note refers, an applicant examined December 1 may be reexamined February 1, and an applicant examined December 29, 30, or 31 may be reexamined the last day of February while one examined February 28 may be reexamined April 28.

§ 13.28 Renewal service requirements, renewal examinations, and exceptions.

A restricted radiotelephone operator permit normally is issued for the lifetime of the holder and need not be renewed, EXCEPT that alien restricted radiotelephone operator permits are normally issued for a five year term and are normally renewable. A license of any other class may be renewed without examination provided that the service record on the reverse side of the license (see §§ 13.91 to 13.94) shows at least two years of satisfactory service in the aggregate during the license term and while actually employed as a radio operator under the license. If this two-year renewal service requirement is not fulfilled, but the service record shows at least one year of satisfactory service in the aggregate during the last three years of the license term and while actually employed as a radio operator under that license, the license may be renewed upon the successful completion of a renewal examination, which may be taken at any time during the final year of the license term or during a one-year period of grace after the date of expiration of the license sought to be renewed. The renewal examination will consist of the highest numbered examination element normally required for a new license of the class sought to be renewed, plus the code test (if any) required for such a new license. If the renewal examination is not successfully completed before expiration of the aforementioned one-year period of grace, the license will not be renewed on any basis.

NOTE: By order dated and effective April 4, 1951, the Commission temporarily waived the requirement of prior service as a radio operator or examination for renewal in the case of any applicant for renewal of his commercial radio operator license. This order is applicable to commercial radio operator licenses which expired after June 30, 1950 until further order of the Commission.

* * * * * * *

SCOPE OF AUTHORITY

§ 13.61 Operating authority.

The various classes of commercial radio operator licenses issued by the Commission authorize the holders thereof to operate radio stations, except amateur, as follows (See also § 13.62(c) for additional operating authority with respect to standard and FM broadcast stations):

(a) *Radiotelegraph first-class operator license.* Any station except:

(1) Stations transmitting television, or

(2) Any of the various classes of broadcast stations, or

(3) On a cargo vessel (other than a vessel operated exclusively on the Great Lakes) required by treaty or statute to be equipped with a radiotelegraph installation, the holder of this class of license may not act as chief or sole operator until he has had at least 6 months' satisfactory service in the aggregate as a qualified radiotelegraph operator in a station on board a ship or ships of the United States.

(4) On an aircraft employing radiotelegraphy, the holder of this class of license may not operate the radiotelegraph station during the course of normal rendition of service unless he has satisfactorily completed a supplementary examination qualifying him for that duty, or unless he has served satisfactorily as chief or sole radio operator on an aircraft employing radiotelegraphy prior to February 15, 1950. The supplementary examination shall consist of:

(1) Written examination element: 7.

(5) At a ship radar station, the holder of this class of license may not supervise or be responsible for the performance of any adjustments or tests during or coincident with the installation, servicing or maintenance of the radar equipment while it is radiating energy unless he has satisfactorily completed a supplementary examination qualifying him for that duty and received a ship radar endorsement on his license certifying to that fact: *Provided,* That nothing in this subparagraph shall be construed to prevent persons holding licenses not so endorsed from making replacements of fuses or of receiving-type tubes. The supplementary examination shall consist of:

(i) Written examination element: 8.

(b) *Radiotelegraph second-class operator license.* Any station except:

(1) Stations transmitting television, or

(2) Any of the various classes of broadcast stations, or

(3) On a passenger vessel (a ship shall be considered a passenger ship if it carries or is licensed or certificated to carry more than 12 passengers; a cargo ship means any ship not a passenger ship) required by treaty or statute to maintain a continuous radio watch by operators or on a vessel having continuous hours of service for public correspondence, the holder of this class of license may not act as chief operator, or

(4) On a vessel (other than a vessel operated exclusively on the Great Lakes) required by treaty or statute to be equipped with a radiotelegraph installation, the holder of this class of license may not act as chief or sole operator until he has had at least 6 months' satisfactory service in the aggregate as a qualified radiotelegraph operator in a station on board a ship or ships of the United States.

(5) On an aircraft employing radiotelegraphy, the holder of this class of license may not operate the radiotelegraph station during the course of normal rendition of service unless he is at least eighteen (18) years of age and has satisfactorily completed a supplementary examination qualifying him for that duty, or unless he has served satisfactorily as chief or sole radio operator on an aircraft employing radiotelegraphy prior to February 15, 1950. The supplementary examination shall consist of:

(i) Transmitting and receiving code test at twenty-five (25) words per minute plain language and twenty (20) code groups per minute.

(ii) Written examination element: 7.

(6) At a ship radar station, the holder of this class of license may not supervise or be responsible for the performance of any adjustments or tests during or coincident with the installation, servicing or maintenance of the radar equipment while it is radiating energy unless he has satisfactorily completed a supplementary examination qualifying him for that duty and received a ship radar endorsement on his license certifying to that fact: *Provided,* That nothing in this subparagraph shall be construed to prevent persons holding licenses not so endorsed from making replacements of fuses or of receiving-type tubes. The supplementary examination shall consist of:

(i) Written examination element: 8.

(c) [Reserved].

(d) *Radiotelegraph third-class operator permit.* Any station except:

(1) Stations transmitting television, or

(2) Any of the various classes of broadcast stations, or

(3) Class I-B coast stations (other than when transmitting manual radiotelegraphy for identification or for testing) at which the power in the antenna of the unmodulated carrier wave is authorized to exceed 250 watts, or

(4) Class II-B or Class III-B coast stations (other than those in Alaska and other than when transmitting manual radiotelegraphy for identification or for testing) at which the power in the antenna of the unmodulated carrier wave is authorized to exceed 250 watts, or

(5) Ship stations or aircraft stations other than those at which the installation is used solely for telephony and at which the power in the antenna of the unmodulated carrier wave is not authorized to exceed 250 watts, or

(6) Ship stations and coast stations open to public correspondence by telegraphy, or

(7) Radiotelegraph stations on board a vessel required by treaty or statute to be equipped with a radio installation, or

(8) Aircraft stations while employing radiotelegraphy:

Provided, That (1) such operator is prohibited from making any adjustments that may result in improper transmitter operation, and (2) the equipment is so designed that the stability of the frequencies of the transmitter is maintained by the transmitter itself within the limits of tolerance specified by the station license, and none of the operations necessary to be performed during the course of normal rendition of the service of the station may cause off-frequency operation or result in any unauthorized radiation, and (3) any needed adjustments of the transmitter that may affect the proper operation of the station are regularly made by or under the immediate supervision and responsibility of a person holding a first- or second-class commercial radio operator license, either radiotelephone or radiotelegraph as may be appropriate for the class of station involved (as determined by the scope of the authority of the respective licenses as set forth in paragraphs (a), (b), (e), and (f) of this section and § 13.62), who shall be responsible for the proper functioning of the station equipment, and (4) in the case of ship radio telephone or aircraft radiotelephone stations when the power in the antenna of the unmodulated carrier wave is authorized to exceed 100 watts, any needed adjustments of the transmitter that may affect the proper operation of the station are made only by or under the immediate supervision and responsibility of an operator holding a first- or second-class radiotelegraph license, who shall be responsible for the proper functioning of the station equipment.

(e) *Radiotelephone first-class operator license.* Any station except:

(1) Stations transmitting telegraphy by any type of the Morse code, or

(2) [Reserved]

(3) At a ship radar station, the holder of this class of license may not supervise or be responsible for the performance of any adjustments or tests during or coincident with the installation, servicing or maintenance of the radar equipment while it is radiating energy unless he has satisfactorily completed a supplementary examination qualifying him for that duty and received a ship radar endorsement on his license certifying to that fact: *Provided,* That nothing in this subparagraph shall be construed to prevent persons holding licenses not so endorsed from making replacements of fuses or of receiving-type tubes. The supplementary examination shall consist of:

(i) Written examination element: 8.

(f) *Radiotelephone second-class operator license.* Any station except:

(1) Stations transmitting telegraphy by any type of the Morse Code, or

(2) Standard broadcast stations, or

(3) International broadcast stations, or

(4) FM broadcast stations, or

(5) Non-commercial educational FM broadcast stations with transmitter power rating in excess of 1 kilowatt, or

(6) Television broadcast stations, or

(7) [Reserved]

(8) At a ship radar station, the holder of this class of license may not supervise or be responsible for the performance of any adjustments or tests during or coincident with the installation, servicing or maintenance of the radar equipment while it is radiating energy unless he has satisfactorily completed a supplementary examination qualifying him for that duty and received a ship radar endorsement on his license certifying to that fact: *Provided,* That nothing in this subparagraph shall be construed to prevent persons holding licenses not so endorsed from making replacements of fuses or of receiving-type tubes. The supplementary examination shall consist of:

(i) Written examination element: 8.

(g) *Radiotelephone third-class operator permit.* Any station except:

(1) Stations transmitting television other than Instructional Television Fixed Service stations, or

(2) Stations transmitting telegraphy by any type of the Morse Code, or

(3) Any of the various classes of broadcast stations, or

(4) Class I–B coast stations at which the power is authorized to exceed 250 watts carrier power or 1,000 watts peak envelope power, or

(5) Class II–B or Class III–B coast stations, other than those in Alaska, at which the power is authorized to exceed 250 watts carrier power or 1,000 watts peak envelope power, or

(6) Ship stations or aircraft stations at which the installation is not used solely for telephony, or at which the power is more than 250 watts carrier power or 1,000 watts peak envelope power:

Provided, That (1) such operator is prohibited from making any adjustments that may result in improper transmitter operation, and (2) the equipment is so designed that the stability of the frequencies of the transmitter is maintained by the transmitter itself within the limits of tolerance specified by the station license, and none of the operations necessary to be performed during the course of normal rendition of the service of the station may cause off-frequency operation or result in any unauthorized radiation, and (3) any needed adjustments of the transmitter that may affect the proper operation of the station are regularly made by or under the immediate supervision and responsibility of a person holding a first- or second-class commercial radio operator license, either radiotelephone or radiotelegraph as may be appropriate for the class of station involved (as determined by the scope of the authority of the respective licenses as set forth in paragraphs (a), (b), (e), and (f) of this section and § 13.62), who shall be responsible for the proper functioning of the station equipment, and (4) in the case of ship radiotelephone or aircraft radiotelephone stations when the power in the antenna of the unmodulated carrier wave is authorized to exceed 100 watts, any needed adjustments of the transmitter that may affect the proper operation of the station are made only by or under the immediate supervision and responsibility of an operator holding a first- or second-class radiotelegraph license, who shall be responsible for the proper functioning of the station equipment.

(h) *Restricted radiotelephone operator permit.* any station except:

(1) Stations transmitting television, or

(2) Stations transmitting telegraphy by any type of the Morse Code, or

(3) Any of the various classes of broadcast stations other than FM translator and booster stations, or

(4) Ship stations licensed to use telephony at which the power is more than 100 watts carrier power or 400 watts peak envelope power, or

(5) Radio stations provided on board vessels for safety purposes pursuant to statute or treaty, or

(6) Coast stations, other than those in Alaska, while employing a frequency below 30 MHz, or

(7) Coast stations at which the power is authorized to exceed 250 watts carrier power or 1,000 watts peak envelope power;

(8) At a ship radar station the holder of this class

of license may not supervise or be responsible for the performance of any adjustments or tests during or coincident with the installation, servicing or maintenance of the radar equipment while it is radiating energy: *Provided,* That nothing in this subparagraph shall be construed to prevent any person holding such a license from making replacements of fuses or of receiving type tubes:

Provided, That, with respect to any station which the holder of this class of license may operate, such operator is prohibited from making any adjustments that may result in improper transmitter operation, and the equipment is so designed that the stability of the frequencies of the transmitter is maintained by the transmitter itself within the limits of tolerance specified by the station license, and none of the operations necessary to be performed during the course of normal rendition of the service of the station may cause off-frequency operation or result in any unauthorized radiation, and any needed adjustments of the transmitter that may affect the proper operation of the station are regularly made by or under the immediate supervision and responsibility of a person holding a first- or second-class commercial radio operator license, either radiotelephone or radiotelegraph, who shall be responsible for the proper functioning of the station equipment.

§ 13.62 Special privileges.

In addition to the operating authority granted under § 13.61, the following special privileges are granted the holders of commercial radio operator licenses:

(a) [Reserved]

(b) The holder of any class of radiotelephone operator's license, whose license authorizes him to operate a station while transmitting telephony, may operate the same station when transmitting on the same frequencies, any type of telegraphy under the following conditions:

(1) When transmitting telegraphy by automatic means for identification, for testing, or for actuating an automatic selective signaling device, or

(2) When properly serving as a relay station and for that purpose retransmitting by automatic means, solely on frequencies above 50 Mc/s, the signals of a radiotelegraph station, or

(3) When transmitting telegraphy as an incidental part of a program intended to be received by the general public, either directly or through the intermediary of a relay station or stations.

(c) The holder of a commercial radiotelegraph first- or second-class license, a radiotelephone second-class license, or a radiotelegraph or radiotelephone third-class permit, endorsed for broadcast station operation may operate any class of standard, FM, or educational FM broadcast station except those using directional antenna systems which are required by the station authorizations to maintain ratios of the currents in the elements of the systems within a tolerance which is less than five percent or relative phases within tolerances which are less than three degrees, under the following conditions:

(1) That adjustments of transmitting equipment by such operators, except when under the immediate supervision of a radiotelephone first-class operator (ra-diotelephone second-class operator for educational FM stations with transmitter output power of 1000 watts or less), and except as provided in paragraph (d) of this section, shall be limited to the following:

(i) Those necessary to turn the transmitter on and off;

(ii) Those necessary to compensate for voltage fluctuations in the primary power supply;

(iii) Those necessary to maintain modulation levels of the transmitter within prescribed limits;

(iv) Those necessary to effect routine changes in operating power which are required by the station authorization;

(v) Those necessary to change between nondirectional and directional or between differing radiation patterns, provided that such changes require only activation of switches and do not involve the manual tuning of the transmitter's final amplifier or antenna phasor equipment. The switching equipment shall be so arranged that the failure of any relay in the directional antenna system to activate properly will cause the emissions of the station to terminate.

(2) The emissions of the station shall be terminated immediately whenever the transmitting system is observed operating beyond the upper and lower limiting values of parameters required to be observed and logged or in any manner inconsistent with the rules or the station authorization, and the above adjustments are ineffective in correcting the condition of improper operation, and a first-class radiotelephone operator is not present.

(3) The special operating authority granted in this section with respect to broadcast stations is subject to the condition that there shall be in employment at the station in accordance with Part 73 of this chapter one or more first-class radiotelephone operators authorized to make or supervise all adjustments, whose primary duty shall be to affect and insure the proper functioning of the transmitting system. In the case of a noncommercial educational FM broadcast station with authorized transmitter output power of 1000 watts or less, a second-class radiotelephone licensed operator may be employed in lieu of a first-class licensed operator.

(d) When an emergency action condition is declared, a person holding any class of radio operator license or permit who is authorized thereunder to perform limited operation of a standard broadcast station may make any adjustments necessary to effect operation in the emergency broadcast system in accordance with the station's National Defense Emergency Authorization: *Provided,* That the station's responsible first-class radiotelephone operator(s) shall have previously instructed such person in the adjustments to the transmitter which are necessary to accomplish operation in the Emergency Broadcast System.

* * * * * * *

§ 13.64 Obedience to lawful orders.

All licensed radio operators shall obey and carry out the lawful orders of the master or person lawfully in charge of the ship or aircraft on which they are employed.

§ 13.65 Damage to apparatus.

No licensed radio operator shall willfully damage, or cause or permit to be damaged, any radio apparatus or installation in any licensed radio station.

§ 13.66 Unnecessary, unidentified, or superfluous communications.

No licensed radio operator shall transmit unnecessary, unidentified, or superfluous radio communications or signals.

§ 13.67 Obscenity, indecency, profanity.

No licensed radio operator or other person shall transmit communications containing obscene, indecent, or profane words, language, or meaning.

§ 13.68 False signals.

No licensed radio operator shall transmit false or deceptive signals or communications by radio, or any call letter or signal which has not been assigned by proper authority to the radio station he is operating.

§ 13.69 Interference.

No licensed radio operator shall willfully or maliciously interfere with or cause interference to any radio communication or signal.

§ 13.70 Fraudulent licenses.

No licensed radio operator or other person shall alter, duplicate, or fraudulently obtain or attempt to obtain, or assist another to alter, duplicate, or fraudulently obtain or attempt to obtain an operator license. Nor shall any person use a license issued to another or a license that he knows to be altered, duplicated, or fraudulently obtained:

MISCELLANEOUS

§ 13.71 Issue of duplicate or replacement licenses.

(a) If the authorization is of the diploma form, a properly executed application for duplicate should be submitted to the office of issue. If the authorization is of the card form (Restricted Radiotelephone Operator Permit), a properly executed application for replacement should be submitted to the Federal Communications Commission, Gettysburg, Pa., 17325, EXCEPT for alien restricted radiotelephone operator permit applications, which must be submitted to Federal Communications Commission, Washington, D.C. 20554. In either case, the application shall embody a statement of the circumstances involved in the loss, mutilation, or destruction of the license or permit. If the authorization has been lost, the applicant must state that reasonable search has been made for it, and further, that in the event it be found, either the original or the duplicate (or replacement) will be returned for cancellation. If the authorization is of the diploma form, the applicant should also submit documentary evidence of the service that has been obtained under the original authorization, or a statement embodying that information.

(b) The holder of any license or permit whose name is legally changed may make application for a replacement document to indicate the new legal name by submitting a properly executed application accompanied by the license or permit affected. If the authorization is of the diploma form, the application should be submitted to the office where it was issued. If the authorization is of the card form (Restricted Radiotelephone Operator Permit) it should be submitted to the Federal Communications Commission, Gettysburg, Pa. 17325, except for alien restricted radiotelephone operator permit applications, which must be submitted to Federal Communications Commission, Washington, D.C. 20554.

NOTE: Pursuant to § 1.1117(c) of this chapter, no fee is required for application for replacement of license for a marriage-related change of name.

§ 13.72 Exhibiting signed copy of application.

When a duplicate or replacement operator license or permit has been requested, or request has been made for renewal, or a request has been made for an endorsement, higher class license or permit, or verification card, the operator shall exhibit in lieu of the original document a signed copy of the application which has been submitted to the Commission.

§ 13.73 Verification card.

The holder of an operator license or permit of the diploma form (as distinguished from such document of the card form) may, by filing a properly executed application accompanied by his license or permit, obtain a verification card (Form 758–F). This card may be carried on the person of the operator in lieu of the original license or permit when operating any station at which posting of an operator license is not required: *Provided*, That the license is readily accessible within a reasonable time for inspection upon demand by an authorized Government representative.

* * * * * *

§ 81.179 Message charges.

(a) (1) No charge shall be made for the service of any public coast station unless effective tariffs applicable to such service are on file with the Commission, pursuant to the requirements of Section 203 of the Communications Act and Part 61 of this chapter.

(2) No charge shall be made for the service of any station subject to this part, other than a public coast station, except as provided by and in accordance with § 81.352.

(b) No charge shall be made by any station in the maritime mobile service of the United States for the transmission of distress messages and replies thereto in connection with situations involving the safey of life and property at sea.

(c) No charge shall be made by any station in the maritime mobile service of the United States for the transmission, receipt, or relay of the information concerning dangers to navigation designated in § 83.303 (b) of this chapter, originating on a ship of the United States or of a foreign country.

* * * * * *

§ 81.302 Points of communication.

(a) Subject to the conditions and limitations imposed by the terms of the particular coast station license or by the applicable provisions of this part with respect to the use of particular radiochannels, public coast stations using telephony are authorized to communicate:

(1) With any ship station or aircraft station operating in the maritime mobile service for the transmission or reception of safety communication;

(2) With any land station for the purpose of facilitating the transmission or reception of safety communication to or from a ship or aircraft station;

* * * * * *

§ 83.6 Operational.

* * * * * *

(f) *Calling.* Transmission from a station solely to secure the attention of another station, or other stations, for a particular purpose.

(g) *Working.* Radiocommunication carried on, for a purpose other than calling, by any station or stations using telegraphy, telephony, or facsimile.

* * * * * *

§ 83.165 Posting of operator authorization.

(a) Except as provided in paragraph (b) of this section, when an operator is required for the operation of a station subject to this part, the original authorization of each such operator while he is employed or designated as radio operator of the station shall be posted in a conspicuous place at the principal location on board ship at which the station is operated.

(b) An operator who holds a Restricted Radiotelephone Operator Permit or a valid license verification card (FCC Form 758–F) attesting to the existence of a commercial radio operator license of the diploma type, may, in lieu of posting, have such permit or verification card in his personal possession immediately available for inspection upon request by a Commission representative when operating the following:

(1) A station which is not required to be installed on the vessel by reason of statute or treaty to which the United States is a party;

(2) Any class of ship station when the operator is on board solely for the purpose of servicing the radio equipment;

(3) A station of a portable nature.

* * * * * * *

§ 83.223 Watch on 2182 kHz.

(a) Each ship station on board a ship navigating the Great Lakes and licensed to transmit by telephony on one or more frequencies within the band 1605 to 3500 kHz shall, during its hours of service for telephony, maintain an efficient watch for reception of A3 and A3H emissions on the authorized carrier frequency 2182 kHz, whenever the station is not being used for transmission on that frequency or for communication on other frequencies.

(b) Except for stations on board vessels required by law to be fitted with radiotelegraph equipment, each ship station (in addition to those ship stations specified in paragraph (a) of this section) licensed to transmit by telephony on one or more frequencies within the band 1605 to 3500 kHz shall, during its hours of service for telephony, maintain an efficient watch for the reception of A3 and A3H emissions on the authorized carrier frequency 2182 kHz, whenever such station is not being used for transmission on that frequency or for communication on other frequencies. When the ship station is in Region 1 or 3, such watch shall, insofar as is possible, be maintained at least twice each hour for 3 minutes commencing at x h. 00 and x h. 30, Greenwich mean time.

* * * * * * *

§ 83.234 Distress signals.

(a) The international radiotelegraph distress signal consists of the group "three dots, three dashes, three dots" (. . . — — — . . .), symbolized herein by $\overline{SOS}$, transmitted as a single signal in which the dashes are slightly prolonged so as to be distinguished clearly from the dots.

(b) The international radiotelephone distress signal consists of the word MAYDAY, pronounced as the French expression "m'aider".

(c) These distress signals indicate that a mobile station is threatened by grave and imminent danger and requests immediate assistance.

§ 83.235 Distress calls.

(a) The distress call sent by radiotelegraphy consists of:
(1) The distress signal $\overline{SOS}$, sent three times;
(2) The word DE;
(3) The call sign of the mobile station in distress, sent three times.

(b) The distress call sent by radiotelephony consists of:
(1) The distress signal MAYDAY spoken three times;
(2) The words THIS IS;
(3) The call sign (or name, if no call sign assigned) of the mobile station in distress, spoken three times.

(c) The distress call shall have absolute priority over all other transmissions. All stations which hear it shall immediately cease any transmission capable of interfering with the distress traffic and shall continue to listen on the frequency used for the emission of the distress call. This call shall not be addressed to a particular station and acknowledgment of receipt shall not be given before the distress message which follows it is sent.

§ 83.236 Distress messages.

(a) The radiotelegraph distress message consists of:
(1) The distress signal $\overline{SOS}$;
(2) The name of the mobile station in distress;
(3) Particulars of its position;
(4) The nature of the distress;
(5) The kind of assistance desired;
(6) Any other information which might facilitate rescue.

(b) The radiotelephone distress message consists of:
(1) The distress signal MAYDAY;
(2) The name of the mobile station in distress;
(3) Particulars of its position;
(4) The nature of the distress;
(5) The kind of assistance desired;
(6) Any other information which might facilitate rescue (for example, the length, color, and type of vessel; number of persons on board, etc.).

(c) As a general rule, a ship shall signal its position in latitude and longitude (Greenwich), using figures for the degrees and minutes, together with one of the words NORTH or SOUTH and one of the words EAST or WEST. In radiotelegraphy, the signal . — . — . — shall be used to separate the degrees from the minutes. When practicable, the true bearing and distance in nautical miles from a known geographical position may be given.

* * * * * * *

§ 83.238 Radiotelephone distress call and message transmission procedure.

(a) The radiotelephone distress procedure shall consist of:
(1) The radiotelephone alarm signal (whenever possible);
(2) The distress call;
(3) The distress message.

(b) The radiotelephone distress transmissions shall be made slowly and distinctly, each word being clearly pronounced to facilitate transcription.

(c) After the transmission by radiotelephony of its distress message, the mobile station may be requested to transmit suitable signals followed by its call sign or name, to permit direction-finding stations to determine its position. This request may be repeated at frequent intervals if necessary.

(d) The distress message, preceded by the distress call, shall be repeated at intervals until an answer is received. This repetition shall be preceded by the radiotelephone alarm signal whenever possible.

(e) When the mobile station in distress receives no answer to a distress message transmitted on the distress frequency, the message may be repeated on any other available frequency on which attention might be attracted.

§ 83.239 Acknowledgment of receipt of distress message.

(a) Stations of the maritime mobile service which receive a distress message from a mobile station which is, beyond any possible doubt, in their vicinity, shall immediately acknowledge receipt. However, in areas where reliable communication with one or more coast stations are practicable, ship stations may defer this acknowledgement for a short interval so that a coast station may acknowledge receipt.

(b) Stations of the maritime mobile service which receive a distress message from a mobile station which, beyond any possible doubt, is not in their vicinity, shall allow a short interval of time to elapse before acknowledging receipt of the message, in order to permit stations nearer to the mobile station in distress to acknowledge receipt without interference.

§ 83.240 Form of acknowledgment.

(a) The acknowledgement of receipt of a distress message is transmitted, when radiotelegraphy is used, in the following form:

(1) The call sign of the station sending the distress message, sent three times;
(2) The word DE;
(3) The call sign of the station acknowledging receipt, sent three times;
(4) The group RRR;
(5) The distress signal $\overline{SOS}$

(b) The acknowledgment of receipt of a distress message is transmitted, when radiotelephony is used, in the following form:

(1) The call sign or other identification of the station sending the distress message, spoken three times;
(2) The words THIS IS;
(3) The call sign or other identification of the station acknowledging receipt, spoken three times;
(4) The word RECEIVED;
(5) The distress signal MAYDAY.

§ 83.241 Information furnished by acknowledging station.

(a) Every mobile station which acknowledges receipt of a distress message shall, on the order of the master or person responsible for the ship, aircraft, or other vehicle carrying such mobile station, transmit as soon as possible the following information in the order shown:

(1) Its name;
(2) Its position, in the f o r m prescribed in § 83.236(c);
(3) The speed at which it is proceeding towards, and the approximate time it will take to reach, the mobile station in distress.

(b) Before sending this message, the station shall ensure that it will not interfere with the emissions of other stations better situated to render immediate assistance to the station in distress.

§ 83.242 Transmission of distress message by a station not itself in distress.

(a) A mobile station or a land station which learns that a mobile station is in distress shall transmit a distress message in any of the following cases:

(1) When the station in distress is not itself in a position to transmit the distress message;
(2) When the master or person responsible for the ship, aircraft, or other vehicle not in distress, or the person responsible for the land station, considers that further help is necessary;
(3) When, although not in a position to render assistance, it has heard a distress message which has not been acknowledged. When a mobile station transmits a distress message under these conditions, it shall take all necessary steps to notify the authorities who may be able to render assistance.

(b) The transmission of a distress message under the conditions prescribed in paragraph (a) of this section shall be made on either or both of the international distress frequencies (500 kHz radiotelegraph; 2182 kHz radiotelephone) or on any other available frequency on which attention might be attracted.

(c) The transmission of the distress message shall always be preceded by the call indicated below, which shall itself be preceded whenever possible by the radiotelegraph or radiotelephone alarm signal. This call consists of:

(1) When radiotelegraphy is used:
(i) the signal $\overline{DDD}$ $\overline{SOS}$ $\overline{SOS}$ $\overline{SOS}$ $\overline{DDD}$;
(ii) The word DE;
(iii) The call sign of the transmitting station, sent three times.

(2) When radiotelephony is used:
(i) The signal MAYDAY RELAY, spoken three times;
(ii) The words THIS IS;
(iii) The call sign or other identification of the trnsmitting station, spoken three times.

(d) When the radiotelegraph alarm signal is used, an interval of two minutes shall be allowed, whenever this is considered necessary, before the transmission of the call mentioned in subparagraph (c)(1) of this section.

§ 83.243 Control of distress traffic.

(a) Distress traffic consists of all messages relating to the immediate assistance required by the mobile station in distress. In distress traffic, the distress signal shall be sent before the call and at the beginning of the preamble of any radiotelegram.

(b) The control of distress traffic is the responsibility of the mobile station in distress or of the station which, pursuant to § 83.242(a), has sent the distress message. These stations may, however, delegate the control of the distress traffic to another station.

(c) The station in distress or the station in control of distress traffic may impose silence either on all stations of the mobile service in the area or on any station which interferes with the distress traffic. It shall address these instructions "to all stations" or to one station only, according to circumstances. In either case, it shall use:

(1) In radiotelegraphy, the abbreviation QRT, followed by the distress signal SOS. The use of the signal QRT SOS shall be reserved for the mobile station in distress and for the station controlling distress traffic;

(2) In radiotelephony, the signal SEELONCE MAYDAY. The use of this signal shall be reserved for the mobile station in distress and for the station controlling distress traffic.

(d) If it is believed to be essential, any station of the mobile service near the ship, aircraft, or other vehicle in distress, may also impose silence. It shall use for this purpose:

(1) In radiotelegraphy, the abbreviation QRT, followed by the word DISTRESS and its own call sign;

(2) In radiotelephony, the word SEELONCE, followed by the word DISTRESS and its own call sign or other identification.

* * * * * * *

§ 83.247 Urgency signals.

(a) The urgency signal indicates that the calling station has a very urgent message to transmit concerning the safety of a ship, aircraft, or other vehicle, or the safety of a person. The urgency signal shall be sent only on the authority of the master or person responsible for the mobile station.

* * * * * *

(c) In radiotelephony, the urgency signal consists of the word PAN, spoken three times and transmitted before the call.

* * * * * * *

§ 83.249 Safety signals.

(a) The safety signal indicates that the station is about to transmit a message concerning the safety of navigation or giving important meteorological warnings.

(b) In radiotelegraphy, the safety signal consists of three repetitions of the group TTT, sent with the individual letters of each group, and the successive groups clearly separated from each other. It shall be sent before the call.

(c) In radiotelephony, the safety signal consists of the word SECURITY, spoken three times and transmitted before the call.

(d) The safety signal and call shall be sent on one of the international distress frequencies (500 kHz radiotelegraph; 2182 kHz radiotelephone), or on the national distress frequency (156.800 MHz radiotelephone). However, stations which cannot transmit on a distress frequency may use any other available frequency on which attention might be attracted.

* * * * * * *

§ 83.352 Frequencies for use in distress.

(a) The frequency 2182 kHz is the international distress frequency for radiotelephony. It shall be used for this purpose by ship, aircraft, and survival craft stations using frequencies in the authorized bands between 1605 and 4000 kHz when requesting assistance from the maritime services.

(b) The frequency 121.5 MHz (using class A2 emission) is available for radio-beacon purposes to survival craft stations. The frequency 121.5 MHz (using A9 emission) is available to emergency position indicating radiobeacon (EPIRB) stations for facilitating search and rescue operations. The frequency 121.5 MHz (class A3 emission) is available to authorized ship stations for emergency communications between ships and aircraft only if operation on the auxiliary frequency 123.1 MHz is also provided for. The frequency 121.5 MHz is available to authorized ship stations for A3 emission for emergency communications between ships and aircraft involved in coordinated search and rescue operations when other VHF channels are not available. As soon as practicable after establishing contact on 121.5 MHz ships and aircraft engaged in the search and rescue operation should shift to the auxiliary frequency 123.1 MHz. The universal auxiliary frequency 123.1 MHz is available to mobile stations engaged in scene of action search and rescue operations as the auxiliary to the emergency clear channel frequency 121.5 MHz.

(c) The frequency 243 MHz (class A9 emission only) is available to EPIRB stations for facilitating search and rescue operations.

§ 83.353 Frequencies for calling.

(a) The international general radiotelephone calling frequency for the maritime mobile service is 2182 kHz. It may be used as a carrier frequency for this purpose by ship stations and aircraft stations operating in the maritime mobile service:

(1) In addition this frequency may be used for transmission of:

(i) The international urgency signal, and very urgent messages (preceded by this signal) concerning the safety of a ship, aircraft, or other vehicle, or the safety of some person on board or within sight of such ship, aircraft, or vehicle.

(ii) The international safety signal, and messages (preceded by this signal) concerning the safety of navigation or giving important meteorological warnings; however, safety messages shall be transmitted, when practicable, on a working frequency after a preliminary announcement on 2182 kHz.

(iii) Brief radio operating signals.

(iv) Brief test signals in accordance with the provisions of § 83.365, as may be necessary to determine whether the radio transmitting equipment of the station is in good working condition on this frequency.

(b) The frequency 156.8 MHz is the international safety and calling frequency for the maritime mobile radiotelephone service in the band 156–174 MHz.

* * * * * *

§ 83.365 Procedure in testing.

(a) Ship stations must use every precaution to insure that, when conducting operational transmitter tests, the emissions of the station will not cause harmful interference. Radiation must be reduced to the lowest practicable value and if feasible shall be entirely suppressed. When radiation is necessary or unavoidable, the testing procedure described below shall be followed:

(1) The licensed radio operator or other person responsible for operation of the transmitting apparatus shall ascertain by careful listening that the test emissions will not be likely to interfere with transmissions in progress; if they are likely to interfere with the working of a coast or aeronautical station in the vicinity of the ship station, the consent of the former station(s) must be obtained before the test emissions occur; (see

required procedures in subparagraphs (2) and (3) of this paragraph following);

(2) The applicable identification of the testing station, followed by the word "test" shall be announced on the radio-channel being used for the test, as a warning that test emissions are about to be made on that frequency;

(3) If, as a result of the announcement prescribed in subparagraph (2) of this paragraph, any station transmits by voice the word "wait", testing shall be suspended. When, after an appropriate interval of time such announcement is repeated and no response is observed, and careful listening indicates that harmful interference should not be caused, the operator shall, if further testing is necessary, proceed as set forth in subparagraphs (4) and (5) of this paragraph;

(4) Testing of transmitters shall, insofar as practicable be confined to working frequencies without two way communications; however, 2182 kHz and 156.8 MHz may be used to contact other ship or coast stations when signal reports are necessary. U.S. Coast Guard stations may be contacted on 2182 kHz for test purposes only when tests are being conducted during inspections by Commission representatives or when qualified radio technicians are installing equipment or correcting deficiencies in the station radiotelephone equipment. In these cases the test shall be identified as 'FCC" or "technical" and logged accordingly;

(5) When further testing is necessary beyond the two "test" announcements specified in subparagraphs (2) and (3) of this paragraph, the operator shall announce the word "testing" followed in the case of a voice transmission test by the count "1, 2, 3, 4, * * * etc." or by test phrases or sentences not in conflict with normal operating signals. The test signals in either case shall have a duration not exceeding 10 seconds. At the conclusion of the test, there shall be voice announcement of the official call sign of the testing station. This test transmission shall not be repeated until a period of at least 1 minute has elapsed; on the frequency 2182 kHz or 156.8 MHz a period of at least 5 minutes shall elapse before the test transmission is repeated.

(b) When testing is conducted on any frequency within the bands 2173.5 to 2190.5 kHz, 156.75 to 156.85, 480 to 510 kHz (survival craft transmitters only), or 8362 to 8366 kHz (survival craft transmitters only), no test transmissions shall occur which are likely to actuate any automatic alarm receiver within range. Survival craft stations shall not be tested on the frequency 500 kHz during the 500 kHz silence periods.

§ 83.366 General radiotelephone operating procedure.

(a) *Calling coast stations.* (1) Use by ship stations of the frequency 2182 kHz for calling coast stations, and for replying to calls from coast stations, is authorized; however, whenever practicable such calls and replies shall be made on the appropriate ship-shore working frequency.

(2) Use by ship stations and marine utility stations on board ship of the frequency 156.8 MHz for calling coast stations and marine utility stations on shore, and for replying to calls from such stations, is authorized; however, whenever practicable such calls and replies shall be made on the appropriate ship-shore working frequency.

(b) *Calling ship stations.* (1) Except when other operating procedure is used to expedite safety communication, ship stations, before transmitting on the intership working frequencies 2003, 2142, 2638, 2738, or 2830 kHz, shall first establish communication with other ship stations by call and reply on 2182 kHz: *Provided,* That calls may be initiated on an intership working frequency when it is known that the called vessel maintains a simultaneous watch on such working frequency and on 2182 kHz.

(2) Except when other operating procedure is used to expedite safety communication, the frequency 156.8 MHz shall be used for call and reply by ship stations and marine utility stations on board ship before establishing communication on either of the intership working frequencies 156.3 or 156.4 MHz.

(c) *Change to working frequency.* After establishing communication with another station by call and reply on 2182 kHz or 156.8 MHz, stations on board ship shall change to an authorized working frequency for the transmission of messages which, under the provisions of this subpart, cannot be transmitted on the respective calling frequencies.

(d) *Authorized use of 2003, 2142, 2638, 2738, and 2830 kHz.* The intership working frequencies 2003, 2142, 2638, 2738, and 2830 kHz shall be used for transmissions by ship stations in accordance with the provisions of §§ 83.176, 83.177, and 83.358.

(e) *Simplex operation only.* All transmission on 2003, 2142, 2638, 2738, and 2830 kHz by two or more stations, engaged in any one exchange of signals or communications, shall take place on only one of these frequencies, i.e., the stations involved shall transmit and receive on the same frequency: *Provided,* That this requirement is waived in the event of emergency when by reason of interference or limitation of equipment single frequency operation cannot be used.

(f) *Limitation on duration of calling.* Calling a particular station shall not continue for more than 30 seconds in each instance. If the called station is not heard to reply, that station shall not again be called until after an interval of 2 minutes. When a station called does not reply to a call sent three times at intervals of 2 minutes, the calling shall cease and shall not be renewed until after an interval of 15 minutes; however, if there is no reason to believe that harmful interference will be caused to other communications in progress, the call sent three times at intervals of 2 minutes may be repeated after a pause of not less than 3 minutes. In event of an emergency involving safety, the provisions of this paragraph shall not apply.

(g) *Limitation on duration of working.* Any one exchange of communications between any two ship stations on 2003, 2142, 2638, 2738, or 2830 kHz, or between a ship station and a limited coast station on 2738 or 2830 kHz, shall not exceed 3 minutes in duration after the two stations have established contact by calling and answering. Subsequent to such exchange of communications, the same two stations shall not again use 2003, 2142, 2638, 2738, or 2830 kHz for communication with each other until 10 minutes have elapsed: *Provided,* That this provision shall in no way limit or delay the transmission of communications concerning the safety of life or property.

(h) *Transmission limitation on 2182 kHz and 156.8 MHz.* Any one exchange of signals by ship stations on 2182 kHz or 156.8 MHz (including calls, replies thereto, and operating signals) shall not exceed 2 minutes: *Provided,* That this time limitation is not applicable to the transmission of distress, alarm, urgency, or safety signals, or to messages preceded by one of these signals.

(i) *Limitation on commercial communication.* On frequencies in the band 156–162 MHz, the exchange of commercial communication shall be limited to the minimum practicable transmission time. In the conduct of ship-shore communication, other than distress, stations on board ship shall comply with instructions given by the limited coast station or marine utility station on shore with which they are communicating, in all matters relative to operating practices and procedures and to the suspension of transmission in order to minimize interference.

(j) *2182 kHz silence periods in Regions 1 and 3.* Transmission by ship or survival craft stations when in Regions 1 and 3 (except in the territorial waters of Japan and the Philippines) is prohibited on any frequency (including 2182 kHz within the band 2173.5 to 2190.5 kHz during each 2182 kHz silence period, i.e., for 3 minutes twice each hour beginning at x h. 00 and x h. 30, Greenwich mean time: *Provided, however,* That this provision is not applicable to the transmission of distress, alarm, urgency, or safety signals, or to messages preceded by one of these signals.

*　　*　　*　　*　　*　　*　　*

APPENDIX IV

Extracts From the FCC Rules and Regulations, Parts 17 and 73

This appendix contains reproductions of portions of Parts 17 and 73 of the FCC Rules and Regulations. Even though the information was current at the time this book was printed, the radio operator should have access to and be familiar with a current copy of the Rules and Regulations, since they are subject to revision.

* * * * * * *

§ 17.25 Specifications for the lighting of antenna structures over 150 feet up to and including 300 feet in height.

(a) Antenna structures over 150 feet, up to and including 200 feet in height above ground, which are required to be lighted as a result of notification to the FAA under § 17.7 and antenna structures over 200 feet, up to and including 300 feet in height above ground, shall be lighted as follows:

(1) There shall be installed at the top of the structure one 300 m/m electric code beacon equipped with two 620- or 700-watt lamps (PS–40 Code Beacon type) both lamps to burn simultaneously, and equipped with aviation red color filters. Where a rod or other construction of not more than 20 feet in height and incapable of supporting this beacon is mounted on top of the structure and it is determined that this additional construction does not permit unobstructed visibility of the code beacon from aircraft at any normal angle of approach, there shall be installed two such beacons positioned so as to insure unobstructed visibility of at least one of the beacons from aircraft at any normal angle of approach. The beacons shall be equipped with a flashing mechanism producing not more than 40 flashes per minute nor less than 12 flashes per minute, with a period of darkness equal to approximately one-half of the luminous period.

(2) At the approximate mid point of the overall height of the tower there shall be installed at least two 116- or 125-watt lamps (A21/TS) enclosed in aviation red obstruction light globes. Each light shall be mounted so as to insure unobstructed visibility of at least one light at each level from aircraft at any normal angle of approach.

(3) All lights shall burn continuously or shall be controlled by a light sensitive device adjusted so that the lights will be turned on at a north sky light intensity level of about 35 foot candles and turned off at a north sky light intensity level of about 58 foot candles.

* * * * * * *

§ 17.47 Inspection of tower lights and associated control equipment.

The licensee of any radio station which has an antenna structure requiring illumination pursuant to the provisions of section 303(q) of the Communications Act of 1934, as amended, as outlined elsewhere in this part:

(a) (1) Shall make an observation of the tower lights at least once each 24 hours either visually or by observing an automatic properly maintained indicator designed to register any failure of such lights, to insure that all such lights are functioning properly as required; or alternatively,

(2) Shall provide and properly maintain an automatic alarm system designed to detect any failure of such lights and to provide indication of such failure to the licensee.

(b) Shall inspect at intervals not to exceed 3 months all automatic or mechanical control devices, indicators, and alarm systems associated with the tower lighting to insure that such apparatus is functioning properly.

§ 17.48 Notification of extinguishment or improper functioning of lights.

The licensee of any radio station which has an antenna structure requiring illumination pursuant to the provisions of section 303(q) of the Communications Act of 1934, as amended, as outlined elsewhere in this part:

(a) Shall report immediately by telephone or telegraph to the nearest Flight Service Station or office of the Federal Aviation Administration any observed or otherwise known extinguishment or improper functioning of any top steady burning light or any flashing obstruction light, regardless of its position on the antenna structure, not corrected within 30 minutes. Such reports shall set forth the condition of the light or lights, the circumstances which caused the failure, and the probable date for restoration of service. Further notification by telephone or telegraph shall be given immediately upon resumption of normal operation of the light or lights.

(b) An extinguishment or improper functioning of a steady burning side intermediate light or lights, shall be corrected as soon as possible, but notification to the FAA of such extinguishment or improper functioning is not required.

§ 17.49 Recording of tower light inspections in the station record.

The licensee of any radio station which has an antenna structure requiring illumination shall make the following entries in the station record of the inspections required by § 17.47.

(a) The time the tower lights are turned on and off each day if manually controlled.

(b) The time the daily check of proper operation of the tower lights was made, if automatic alarm system is not provided.

(c) In the event of any observed or otherwise known extinguishment or improper functioning of a tower light:

(1) Nature of such extinguishment or improper functioning.

(2) Date and time the extinguishment or improper functioning was observed, or otherwise noted.

(3) Date, time, and nature of the adjustments, repairs, or replacements made.

(4) Identification of Flight Service Station (Federal Aviation Administration) notified of the extinguishment or improper functioning of any code or rotating beacon light or top light not corrected within 30 minutes, and the date and time such notice was given.

(5) Date and time notice was given to the Flight Service Station (Federal Aviation Administration) that the required illumination was resumed.

(d) Upon completion of the periodic inspection required at least once each 3 months:

(1) The date of the inspection and the condition of all tower lights and associated tower lighting control devices, indicators and alarm systems.

(2) Any adjustments, replacements, or repairs made to insure compliance with the lighting requirements and the date such adjustments, replacements or repairs were made.

* * * * * * *

SUBPART A—STANDARD BROADCAST STATIONS

DEFINITIONS

§ 73.1 Standard broadcast station.

The term "standard broadcast station" means a broadcasting station licensed for the transmission of radiotelephone emissions primarily intended to be received by the general public and operated on a channel in the band 535–1605 kilohertz (kHz).

§ 73.2 Standard broadcast band.

The term "standard broadcast band" means the band of frequencies extending from 535 to 1605 kHz.

§ 73.3 Standard broadcast channel.

The term "standard broadcast channel" means the band of frequencies occupied by the carrier and two side bands of a broadcast signal with the carrier frequency at the center. Channels shall be designated by their assigned carrier frequencies. The 107 carrier frequencies assigned to standard broadcast stations shall begin at 540 kHz and be in successive steps of 10 kHz.

§ 73.4 Dominant station.

The term "dominant station" means a Class I station, as defined in § 73.21, operating on a clear channel.

§ 73.5 Secondary station.

The term "secondary station" means any station, except a Class I station, operating on a clear channel.

§ 73.6 Daytime.

The term "daytime" means that period of time between local sunrise and local sunset.

§ 73.7 Nighttime.

The term "nighttime" means that period of time between local sunset and local sunrise.

§ 73.8 Sunrise and sunset.

The terms "sunrise" and "sunset" mean, for each particular location and during any particular month, the time of sunrise and sunset as specified in the instrument of authorization (See § 73.83).

§ 73.9 Broadcast day.

The term "broadcast day" means that period of time between local sunrise and 12 midnight local time.

§ 73.10 Experimental period.

The term "experimental period" means that time between 12 midnight local time and local sunrise. This period may be used for experimental purposes in testing and maintaining apparatus by the licensee of any standard broadcast station on its assigned frequency and with its authorized power, provided no interference is caused to other stations maintaining a regular operating schedule within such period. No station licensed for "daytime" or "specified hours" of operation may broadcast any regular or scheduled program during this period.

§ 73.11 Service areas.

(a) The term "primary service area" of a broadcast station means the area in which the groundwave is not subject to objectionable interference or objectionable fading.

(b) The term "secondary service area" of a broadcast station means the area served by the skywave and not subject to objectionable interference. The signal is subject to intermittent variations in intensity.

(c) The term "intermittent service area" of a broadcast station means the area receiving service from the groundwave but beyond the primary service area and subject to some interference and fading.

§ 73.12 Portable transmitter.

The term "portable transmitter" means a transmitter so constructed that it may be moved about conveniently from place to place, and is in fact so moved about from time to time, but not ordinarily used while in motion. In the standard broadcast band, such a transmitter is used in making field intensity measurements for locating a transmitter site for a standard broadcast station. A portable broadcast station will not be licensed in the standard broadcast band for regular transmission of programs intended to be received by the public.

§ 73.14 Technical definitions.

(a) *Combined audio harmonics.* The term "combined audio harmonics" means the arithmetical sum of the amplitudes of all the separate harmonic components. Root sum square harmonic readings may be accepted under conditions prescribed by the Commission.

(b) *Effective field.* The term "effective field" or "effective field intensity" is the root-mean-square (RMS) value of the inverse distance fields at a distance of 1 mile from the antenna in all directions in the horizontal plane.

(c) *Nominal power.* "Nominal power" is the power of a standard broadcast station, as specified in a system of classification which includes the following values: 50 kW, 25 kW, 10 kW, 5kW, 2.5 kW, 1 kW, 0.5 kW, 0.25 kW.

(d) *Operating power.* Depending on the context within which it is employed, the term "operating power" may be synon mous with "nominal power" or "antenna power."

(e) *Maximum rated carrier power.* "Maximum rated carrier power" is the maximum power at which the transmitter can be operated satisfactorily and is determined by the design of the transmitter and the type and number of vacuum tubes used in the last radio stage.

(f) *Plate input power.* "Plate input power" means the product of the direct plate voltage applied to the tubes in the last radio stage and the total direct current flowing to the plates of these tubes, measured without modulation.

(g) *Antenna power.* "Antenna input power" or "antenna power" means the product of the square of the antenna current and the antenna resistance at the point where the current is measured.

(h) *Antenna current.* "Antenna current" means the radio-frequency current in the antenna with no modulation.

(i) *Antenna resistance.* "Antenna resistance" means the total resistance of the transmitting antenna system at the operating frequency and at the point at which the antenna current is measured.

(j) *Modulator stage.* "Modulator stage" means the last amplifier stage of the modulating wave which modulates a radio-frequency stage.

(k) *Modulated stage.* "Modulated stage" means the radio-frequency stage to which the modulator is coupled and in which the continuous wave (carrier wave) is modulated in accordance with the system of modulation and the characteristics of the modulating wave.

(l) *Last radio stage.* "Last radio stage" means the oscillator or radio-frequency-power amplifier stage which supplies power to the antenna.

(m) *Percentage modulation* (amplitude):

In a positive direction:

$$M = \frac{MAX - C}{C} \times 100$$

In a negative direction:

$$M = \frac{C - MIN}{C} \times 100$$

Where:
M = Modulation level in percent.
MAX = Instantaneous maximum level of the modulated radio frequency envelope.
MIN = Instantaneous minimum level of the modulated radio frequency envelope.
C = (Carrier) level of radio frequency envelope without modulation.

(n) *Maximum percentage of modulation.* "Maximum percentage of modulation" means the greatest percentage of modulation that may be obtained by a transmitter without producing in its output harmonics of the modulating frequency in excess of those permitted by these regulations.

(o) *High level modulation.* "High level modulation" is modulation produced in the plate circuit of the last radio stage of the system.

(p) *Low level modulation.* "Low level modulation" is modulation produced in an earlier stage than the final.

(q) *Plate modulation.* "Plate modulation" is modulation produced by introduction of the modulating wave into the plate circuit of any tube in which the carrier frequency wave is present.

(r) *Grid modulation.* "Grid modulation" is modulation produced by introduction of the modulating wave into any of the grid circuits of any tube in which the carrier frequency wave is present.

(s) *Blanketing.* Blanketing is that form of interference which is caused by the presence of a broadcast signal of one volt per meter (v/m) or greater intensity in the area adjacent to the antenna of the transmitting station. The 1 v/m contour is referred to as the blanket contour and the area within this contour is referred to as the blanket area.

* * * * * * *

§ 73.39 Indicating instruments—specifications.

(a) Instruments indicating the plate current or plate voltage of the last radio stage (linear scale instruments) shall meet the following specifications:

(1) Length of scale shall be not less than 2 3/10 inches.

(2) Accuracy shall be at least 2 percent of the full scale reading.

(3) The maximum rating of the meter shall be such that it does not read off scale during modulation.

(4) Scale shall have at least 40 divisions.

(5) Full scale reading shall not be greater than five times the minimum normal indication.

(b) Instruments indicating antenna current, common point current, and base currents shall meet the following specifications:

(1) Instruments having logarithmic or square law scales:

(i) Shall meet the requirements of paragraph (a) (1), (2), and (3) of this section for linear scale instruments.

(ii) Full scale reading shall not be greater than three times the minimum normal indication.

(iii) No scale division above one-third full scale reading (in amperes) shall be greater than one-thirtieth of the full scale reading. (Example: An ammeter meeting requirement (i) having full scale reading of 6 amperes is acceptable for reading currents from 2 to 6 amperes, provided no scale division between 2 and 6 amperes is greater than one-thirtieth of 6 amperes, 0.2 ampere.)

(2) Radio frequency instruments having expanded scales:

(i) Shall meet the requirements of paragraph (a) (1), (2), and (3) of this section for linear scale instruments.

(ii) Full scale reading shall not be greater than five times the minimum normal indication.

(iii) No scale division above one-fifth full scale reading (in amperes) shall be greater than one-fiftieth of the full scale reading. (Example: An ammeter meeting the requirement (i) is acceptable for indicating currents from 1 to 5 amperes, provided no division between 1 and 5 amperes is greater than one-fiftieth of 5 amperes, 0.1 ampere.)

(iv) Manufacturers of instruments of the expanded scale type must submit data to the Commission showing that these instruments have acceptable expanded scales, and the type number of these instruments must include suitable designation.

(c) A thermocouple type ammeter, or other device capable of providing an indication of radio frequency current, meeting requirements of paragraph (b) of this section, shall be permanently installed in the antenna circuit or a suitable jack and plug arrangement may be made to permit removal of the meter from the antenna circuit so as to protect it from damage by lightning. Where a jack and plug arrangement is used, contacts shall be made of silver and capable of operating without arcing or heating, and shall be protected against corrosion. Insertion and removal of the meter shall not interrupt the transmissions of the station. When removed from the antenna circuit, the meter shall be labelled to clearly identify the tower in which it is used, and shall be stored in a location which is readily available to that tower. Care shall be exercised in handling the meter to prevent damage which would impair its accuracy. Where the meter is permanently connected in the antenna circuit, provision may be made to short or open the meter circuit when it is not being used to measure antenna current. Such switching shall be accomplished without interrupting the transmission of the station.

(d) Remote reading antenna ammeter(s) may be employed and the indications logged as the antenna current, or in the case of a directional antenna, the common point current and base currents, in accordance with the following:

(1) Remote reading antenna, common point or base ammeters may be provided by:

(i) Inserting second thermocouple directly in the antenna circuit with remote leads to the indicating instrument.

(ii) Inductive coupling to thermocouple or other device for providing direct current to indicating instrument.

(iii) Capacity coupling to thermocouple or other device for providing direct current to indicating instrument.

(iv) Current transformer connected to second thermocouple or other device for providing direct current to indicating instrument.

(v) Using transmission line current meter at transmitter as remote reading ammeter. See subparagraph (7) of this paragraph.

(vi) Using indications of phase monitor for determining the antenna base currents or their ratio in the case of directional antennas, provided that the base current readings are read and logged in accordance with the provision of the station license, and provided further that the indicating instruments in the unit are connected directly in the current sampling circuits with no other shunt circuits of any nature. The meters in the phase monitor may utilize arbitrary scale divisions provided a calibration curve showing the relationship between the arbitrary scale and the scale of the base meters is maintained at the transmitter location.

(vii) Using indications of remote control equipment, provided that the indicating instruments are capable of being connected directly into the antenna circuit at the same point as, but below (transmitter side), the antenna ammeter. The meter(s) in the remote control equipment may utilize an arbitrary scale division provided a calibration curve showing the relationship between the arbitrary scale and the scale of the antenna ammeter is maintained at the remote control point. The meter(s) in the remote control equipment must be calibrated once a week against the regular meter and the results thereof entered in the maintenance log.

(2) Remote ammeters shall be connected into the antenna circuit, at the same point as, but below (transmitter side), the antenna ammeter(s), and shall be calibrated to indicate within 2 percent of the regular meter over the entire range above one-third or one-fifth full scale. See paragraphs (b) (1) (i), (iii) and (b) (2) (i), (iii) of this section.

(3) The regular antenna ammeter, common point ammeter, or base current ammeters shall be above (antenna side) the coupling to the remote meters in the antenna circuit so they do not read the current to ground through the remote meter(s).

(4) All remote meters shall meet the same requirements as the regular antenna ammeter with respect to scale accuracy, etc.

(5) Calibration shall be checked against the regular meter at least once a week.

(6) All remote meters shall be provided with shielding or filters as necessary to prevent any feed-back from the antenna to the transmitter.

(7) In the case of shunt-excited antennas, the transmission line current meter at the transmitter may be considered as the remote antenna ammeter provided the transmission line is terminated directly into the excitation circuit feed line, which shall employ series tuning only (no shunt circuits of any type shall be employed) and insofar as practicable, the type and scale of the transmission line meter should be the same as those of the excitation circuit feed line meter (meter in slant wire feed line or equivalent).

(8) Remote reading antenna ammeters employing vacuum tube rectifiers or semi-conductor devices are acceptable, provided:

(i) The indicating instruments shall meet all the above requirements for linear scale instruments.

(ii) Data are submitted under oath showing the unit has an over-all accuracy of at least 2 percent of the full scale reading.

(iii) The installation, calibration, and checking are in accordance with the requirements of this paragraph.

(9) In the event there is any question as to the method of providing the remote indication, or the accuracy of the remote meter, the burden of proof of satisfactory performance shall be upon the licensee and the manufacturer of the equipment.

(e) [Reserved]

(f) No instrument, the seal of which has been broken, or the accuracy of which is questionable, shall be employed. Any instrument which was not originally sealed by the manufacturer that has been opened shall not be used until it has been recalibrated and sealed in accordance with the following: Repairs and recalibration of instruments shall be made by the manufacturer, by an authorized instrument repair service of the manufacturer, or by some other properly qualified and equipped instrument repair service. In any event, the instrument must be resealed with the symbol or trade-mark of the repair service and a certificate of calibration supplied therewith.

(g) Since it is usually impractical to measure the actual antenna current of a shunt excited antenna system, the current measured at the input of the excitation circuit feed line is accepted as the antenna current.

(h) [Reserved]

(i) The function of each instrument shall be clearly and permanently shown on the instrument itself or on the panel immediately adjacent thereto.

(j) Digital meters, printers, or other numerical readout devices may be used in addition to or in lieu of indicating instruments meeting the specifications of paragraphs (a) and (b) of this section. If a single digital device is used at the transmitter for reading and logging of operating parameters, either (1) indicating instruments meeting the above-mentioned specifications shall be installed in the transmitter and antenna circuit, or (2) a spare digital device shall be maintained at the transmitter with provision for its rapid substitution for the main device should that device malfunction. The readout of the device shall include at least three digits and shall indicate the value or a decimal multiple of the value of the parameter being read to an accuracy of at least 2 percent. The multiplier to be applied to the reading of each parameter shall be indicated at the operating position of a switch used to select the parameter for display, or on the face of an automatically printed log at least once for each calendar day.

*　　*　　*　　*　　*　　*　　*

§ 73.50 Requirements for approval of modulation monitors.

(a) Any manufacturer desiring to submit a monitor for type approval shall supply the Commission with full specification details (two sworn copies) specified in paragraph (b) of this section. If this information appears to meet the requirements of the rules, shipping instructions will be issued to the manufacturer. The shipping charges to and from the Laboratory at Laurel, Maryland, shall be paid for by the manufacturer. Approval of a monitor will only be given on the basis of the data obtained from the sample monitor submitted to the Commission for test.

(1) In approving a monitor upon the basis of the tests conducted by the Laboratory, the Commission merely recognizes that the type of monitor has the inherent capability of functioning in compliance with the rules, if properly constructed, maintained, and operated.

(2) Additional rules with respect to withdrawal of type approval, modification of type approval equipment and limitations on the findings upon which type approval is based are set forth in Part 2, Subpart F, of this chapter.

(b) The specifications that the modulation monitor shall meet before it will be approved by the Commission are as follows:

(1) A DC meter for setting the average rectified carrier at a specific value and to indicate changes in carrier intensity during modulation.

(2) A peak indicating light or similar device that can be set at any predetermined value from 50 to 120 percent modulation to indicate on positive peaks, and/or from 50 to 100 percent negative modulation.

(3) A semi-peak indicator with a meter having the characteristics given below shall be used with a circuit such that peaks of modulation of duration between 40 and 90 milliseconds are indicated to 90 percent of full value and the discharge rate adjusted so that the pointer returns from full reading to 10 percent of zero within 500 to 800 milliseconds. A switch shall be provided so that this meter will read either positive or negative modulation and, if desired, in the center position it may read both in a full-wave circuit. The characteristics of the indicating meter are as follows:

(i) The damping factor shall be between 16 and 200. The useful scale length shall be at least 2.3 inches. The meter shall be calibrated for modulation from 0 to 110 percent and in decibels below 100 percent with 100 percent being 0 dB.

(ii) The accuracy of the reading on percentage of modulation shall be ±2 percent for 100 percent modulation, and ±4 percent of full scale reading at any other percentage of modulation.

(4) The frequency characteristics curve shall not depart from a straight line more than ±½ dB from 30 to 10000 Hz. The amplitude distortion or generation

of audio harmonics shall be kept to a minimum.

(5) The modulation meter shall be equipped with appropriate terminals so that an external peak counter can be readily connected.

(6) Modulation will be tested at 115 volts ±5 percent and 60 Hz, and the above accuracies shall be applicable under these conditions.

(7) All specifications not already covered above and the general design, construction, and operation of these units must be in accordance with good engineering practice.

(c) The modulation monitor may be a part of the frequency monitor.

§ 73.51 Antenna input power; how determined.

(a) Except in those circumstances described in paragraph (d) of this section, the antenna input power shall be determined by the direct method, i.e., as the product of the antenna resistance at the operating frequency (see § 73.54) and the square of the unmodulated antenna current at that frequency, measured at the point where the antenna resistance has been determined.

(b) The authorized antenna input power for each station shall be equal to the nominal power for such station, with the following exceptions:

(1) For stations with nominal powers of 5 kilowatts, or less, the authorized antenna input power to directional antennas shall exceed the nominal power by 8 percent.

(2) For stations with nominal powers in excess of 5 kilowatts, the authorized antenna input power to directional antennas shall exceed the nominal power by 5.3 percent.

(3) In specific cases, it may be necessary to limit the radiated field to a level below that which would result if nominal power were delivered to the antenna. In such cases, excess power may be dissipated in the antenna feed circuit (see § 73.54 (a) and (d)), and/or the transmitter may be operated with power output at a level which is less than the nominal value.

(i) Where a dissipative network is employed, the authorized antenna current and resistance, and the authorized antenna input power shall be determined at the input terminals of the dissipative network.

(ii) Where the authorized antenna input power is less than the nominal power, subject to the conditions set forth in paragraph (c) of this section, the transmitter may be operated at the reduced level necessary to supply the authorized antenna input power.

(c) Applications for authority to operate with antenna input power which is less than nominal power and/or to employ a dissipative network in the antenna system shall be made on FCC Form 302. The technical information supplied on section II–A of this form shall be that applying to the proposed conditions of operation. In addition, the following information shall be furnished, as pertinent.

(1) Full details of any network employed for the purpose of dissipating radio frequency energy otherwise delivered to the antenna (see § 73.54).

(2) A showing that the transmitter has been type accepted for operation at the proposed power output level, or, in lieu thereof:

(i) A full description of the means by which transmitter output power will be reduced.

(ii) Where the proposed transmitter power output level(s) is less than 90 percent of nominal power, equipment performance measurements, as specified in § 73.47 conducted at each proposed power output level; in addition, the measurements and observations required by § 73.47(a) (1), (2), (3) and (5) for power output level 10 percent above, and 10 percent below, the proposed output level(s), but at a modulation level of 95 to 100 percent only. Such measurements must demonstrate that, operating at the proposed power output level(s) the transmitter meets the performance requirements of § 73.40.

(iii) A showing that, at the proposed power output level, means are provided for varying the transmitter output within a tolerance of ±10 percent, to compensate for variations in line voltage or other factors which may affect the power output level.

(d) The antenna input power shall be determined on a temporary basis by the indirect method described in paragraphs (e) and (f) of this section in the following circumstances: (1) In an emergency, where the authorized antenna system has been damaged by causes beyond the control of the licensee or permittee (see § 73.45), or (2) pending completion of authorized changes in the antenna system, or (3) if changes occur in the antenna system or its environment which affect or appear likely to affect the value of antenna resistance, or (4) if the antenna current meter or common point current meter becomes defective, and the static does not employ a remote reading antenna or common point meter (see § 73.58). Prior authorization for the indirect determination of antenna input power is not required. However, an appropriate notation shall be made in the operating log.

(e) (1) Antenna input power is determined indirectly by applying an appropriate factor to the plate input power, in accordance with the following formula:

$$\text{Antenna input power} = Ep \times Ip \times F$$

Where:
Ep = Plate voltage of final radio stage.
Ip = Total plate current of final radio stage.
F = Efficiency factor.

(2) The value of F applicable to each mode of operation shall be entered in the operating log for each date of operation, with a notation as to its derivation. The factor shall be established by one of the methods described in paragraph (f) of this section, which are listed in order of preference. The product of the plate current and plate voltage, or, alternatively, the antenna input power, as determined pursuant to subparagraph (1) of this paragraph, shall be entered in the operating log under an appropriate heading for each log entry plate current and plate voltage.

(f) (1) If the transmitter and the antenna input power utilized during the period of indirect power determination are the same as have been authorized and utilized for any period of regular operation, the factor F shall be the ratio of such authorized antenna input power to the corresponding plate input power of the transmitter for regular conditions of operation, computed with values of plate voltage and plate current obtained from the operating logs of the station for the last week of regular operation.

(2) If a station has not been previously in regular operation with the power authorized for the period indirect power determination, if a new transmitter has been installed, or if, for any other reason, the determination of the factor F by the method described

subparagraph (1) of this paragraph is impracticable:

(i) The factor F shall be obtained from the transmitter manufacturer's letter or test report retained in the station's files, if such a letter or test report specifies a unique value of F for the power level and frequency utilized; or

(ii) By reference to the following table:

Factor (F)	Method of modulation	Maximum rated carrier power	Class of amplifier
0.70	Plate	0.25 to 1.0 kW	
.80	Plate	2.5 kW. and over	
.35	Low level	0.25 kW. and over	B
.65	Low level	0.25 kW. and over	BC [1]
.35	Grid	0.25 kW. and over	

[1] All linear amplifier operation where efficiency approaches that of Class C operation.

NOTE: When the factor F is obtained from the table, this value shall be used even though the antenna input power may be less than the maximum rated carrier power of the transmitter.

(iii) If a station has been authorized to operate with antenna input power which is lower than nominal power, the factor F shall have the value established when such operation was authorized.

§ 73.52 Antenna input power; maintenance of.

(a) The actual antenna input power of each station shall be maintained as near as is practicable to the authorized antenna input power and shall not be less than 90 percent nor greater than 105 percent of the authorized power; except that if, in an emergency, it becomes technically impossible to operate with the authorized power, the station may be operated with reduced power for a period of not more than 30 days without further authority from the Commission, *Provided*, That notification is sent to the Commission in Washington, D.C. not less than the 10th day of the lower power operation. In the event normal power is restored prior to the expiration of the 30 day period, the permittee or licensee will so notify the Commission in Washington, D.C. of this date. If causes beyond the control of the permittee or licensee prevent restoration of authorized power within the allowed period, informal written request shall be made to the Commission in Washington, D.C. no later than the 30th day for such additional time as may be deemed necessary.

(b) In addition to maintaining antenna input power within the above limitations, each station employing a directional antenna shall maintain the relative amplitudes of the antenna currents in the elements of its array within 5 percent of the ratios specified in its license or other instrument of authorization, unless more stringent limits are specified therein.

§ 73.53 Requirements for type approval of antenna monitors.

(a) General requirements:

(1) Any manufacturer desiring to submit a monitor for type approval shall submit an application to the Commission in accordance with the procedure set forth in § 2.561 of this chapter, and subject to the fee schedule in § 1.1120 of this chapter.

(2) Type approval of a monitor is granted subject to the limitations and requirements described in Part 2, Subparts F and I of this chapter.

(b) The Laboratory Division of the Commission will make all tests necessary to determine whether or not the specifications of paragraph (c) of this section have been met. During these tests, the monitor will be operated, to the extent possible, under service conditions. The manufacturer shall furnish to the Laboratory all instructions and services which will be supplied to a purchaser of the monitor.

(c) An antenna monitor eligible for type approval by the Commission shall meet the following specifications:

(1) The monitor shall be designed to operate on a frequency in the band 540 to 1600 kHz.

(2) The monitor shall be capable of indicating any phase difference between two RF voltages of the same frequency over a range of from 0 to 360°.

(3) The monitor shall be capable of indicating the relative amplitude of two RF voltages.

(4) The device used to indicate phase differences shall indicate in degrees, and shall be graduated in increments of 2°, or less. If a digital indicator is provided, the smallest increment shall be 0.5°, or less.

(5) The device used to indicate relative amplitudes shall be graduated in increments which are 1 percent, or less, of the full scale value. If a digital indicator is provided, the smallest increment shall be 0.1 percent, or less, of the full scale value.

(6) The monitor shall be equipped with means, if necessary, to resolve ambiguities in indication.

(7) If the monitor is provided with more than one RF input terminal in addition to a reference input terminal, appropriate switching shall be provided in the monitor so that the signal at each of these RF inputs may be selected separately for comparison with the reference input signal.

(8) Each RF input of the monitor shall provide a termination of such characteristics that, when connected to a sampling line of an impedance specified by the manufacturer, the voltage reflection coefficient shall be 3 percent or less.

(9) The monitor shall be designed so that the switching function required by paragraph (c)(7) of this section may be performed from a point external to the monitor and phase and amplitude indications be provided by external meters. The indications of external meters furnished by the manufacturer shall meet the specifications for accuracy and repeatability of the monitor itself, and the connection of these meters to the monitor, or of other indicating instruments with electrical characteristics meeting the specifications of the monitor manufacturer shall not affect adversely the performance of the monitor in any respect.

(10) If the monitor is fitted with operational features not specifically required by this section, the features:

(i) Shall be arranged so as not to interfere with or be confused with the required functions of the monitor.

(ii) Shall meet the manufacturer's specifications for such operational features.

(11) The monitor shall be accompanied by complete and correct schematic diagrams and operating instructions. For the purpose of type approval, these diagrams and instructions shall be considered as part of the monitor.

(12) The general design, construction and operation of the monitor shall be in accordance with good engineering practice.

(13) When an RF signal of an amplitude within a range specified by the manufacturer is applied to the reference RF input terminal of the monitor, and another RF signal of the same frequency and of equal or lower amplitude is applied to any other selected RF

input terminal, indications shall be provided meeting the following specifications.

(i) The accuracy with which any difference in the phases of the applied signals is indicated shall be ±1°, or better, for signal amplitude ratios of from 2:1 to 1:1, and ±2°, or better, for signal amplitude ratios in excess of 2:1 and up to 5:1.

(ii) The repeatability of indication of any difference in the phases of the applied signals shall be ±1°, or better.

(iii) The accuracy with which the relative amplitudes of the applied signals is indicated, over a range in which the ratio of these amplitudes is between 2:1 and 1:1, shall be ±2 percent of the amplitude ratio, or better, and for amplitude ratios in excess of 2:1 and up to 5:1, ±5 percent of the ratio, or better.

(iv) The repeatability of indication of the relative amplitudes of the applied signals, over a range where the ratio of these amplitudes is between 5:1 and 1:1, shall be ±2 percent of the amplitude ratio, or better.

(v) The modulation of the RF signals by a sinusoidal wave of any frequency between 100 and 10,000 Hz, at any amplitude up to 90 percent shall cause no deviation in an indicated phase difference from its value, as determined without modulation, greater than ±0.5°.

(14) The performance specifications set forth in paragraph (c) (13 of this section, shall be met when the monitor is operated and tested under the following conditions.

(i) After continuous operation for 1 hour, the monitor shall be calibrated and adjusted in accordance with the manufacturer's instructions.

(ii) The monitor shall be subjected to variations in ambient temperature between the limits of 10 and 40°C; external meters furnished by the manufacturer will be subjected to variations between 15 and 30°C.

(iii) Powerline supply voltage shall be varied over a range of from 10 percent below to 10 percent above the rated supply voltage.

(iv) The amplitude of the reference signal shall be varied over the operating range specified by the manufacturer, and in any case over a range of maximum to minimum values of 3 to 1.

(v) The amplitude of the comparison signal shall be varied from a value which is 0.2 of the amplitude of the reference signal to a value which is equal in amplitude to the reference signal.

(vi) Accuracy shall be determined for the most adverse combination of conditions set forth above.

(vii) Repeatability shall be determined as that which may be achieved under the specified test conditions over a period of 7 days, during which no calibration or aljustment of the instrument, subsequent to the initial calibration, shall be made.

(viii) The effects of modulation of the RF signal shall be separately determined, and shall not be included in establishing values for accuracy and repeatability.

* * * * * *

§ 73.55 **Modulation.**

The percentage of modulation shall be maintained at as high a level as is consistent with good quality of transmission and good broadcast service. In no case shall it exceed 100 percent on negative peaks of frequent recurrence, or 125 percent on positive peaks at any time. Generally, modulation should not be less than 85 percent on peaks of frequent recurrence, but when such action may be required to avoid objectionable loudness, the degree of modulation may be reduced to whatever level is necessary for this purpose, even though, under such circumstances, the level may be substantially less than that which produces peaks of frequent recurrence at a level of 85 percent.

§ 73.56 **Modulation monitors.**

(a) Each station shall have in operation, either a the transmitter or the extension meter location, or the place the transmitter is controlled, a modulation monitor of a type approved by the Commission.

NOTE.—Approved modulation monitors are included on the Commission's "Radio Equipment List". Copies of this list are available for inspection at the Commission's offices in Washington, D.C., and at each of its field offices.

(b) In the event that the modulation monitor becomes defective, the station may be operated without the monitor pending its repair or replacement for period not in excess of 60 days without further authority of the Commission: *Provided*, That:

(1) Appropriate entries shall be made in the maintenance log of the station showing the date and time the monitor was removed and restored to service.

(2) The degree of modulation of the station shall be monitored with a cathode ray oscilloscope or other acceptable means.

(c) If conditions beyond the control of the licensee prevent the restoration of the monitor to service within the above allowed period, informal request in accordance with § 1.549 of this chapter may be filed with the Engineer in Charge of the radio district in which the station is operating for such additional time as may be required to complete repairs of the defective instrument.

(d) Each station operated by remote control shall continuously, except when other readings are being taken, monitor percent of modulation or shall be equipped with an automatic device to limit percent of modulation on negative peaks to 100.

* * * * * *

§ 73.58 **Indicating instruments.**

(a) Each standard broadcast station shall be equipped with indicating instruments, which conform with the specifications set forth in § 73.39, for measuring the DC plate circuit current and voltage of the last radio frequency amplified stage; the radio frequency base current of each antenna element; and, for stations employing directional antenna systems, the radio frequency current at the point of common input to the directional antenna.

(b) In the event that any one of these indicating instruments becomes defective when no substitute which conforms with the required specifications is available, the station may be operated without the defective instrument pending its repair or replacement for a period not in excess of 60 days without further authority of the Commission: *Provided*, That:

(1) Appropriate entries shall be made in the maintenance log of the station showing the date and time the meter was removed from and restored to service.

(2) [Reserved]

(3) If the defective instrument is the antenna current meter of a nondirectional station which does

not employ a remote antenna ammeter, or if the defective instrument is the common point meter of a station which employs a directional antenna and does not employ a remote common point meter, the operating power shall be determined by the indirect method in accordance with § 73.51 (d), (e), and (f) during the entire time the station is operated without the antenna current meter or common point meter. However, if a remote antenna ammeter or a remote common point meter is employed and the antenna current meter or common point meter becomes defective, the remote meter may be used in determining operating power by the direct method pending the return to service of the regular meter, provided other meters are maintained at same value previously employed.

(c) If conditions beyond the control of the licensee prevent the restoration of the meter to service within the above allowed period, informal request in accordance with § 1.549 of this chapter may be filed with the Engineer in Charge of the radio district in which the station is located for such additional time as may be required to complete repairs of the defective instrument.

(d) Remote antenna ammeters and remote common point meters are not required; therefore, authority to operate without them is not necessary. However, if a remote antenna ammeter or common point meter is employed and becomes defective, the antenna base currents may be read and logged once daily for each mode of operation, pending the return to service of the regular remote meter.

§ 73.59 Frequency tolerance.

The operating frequency of each station shall be maintained within 20 hertz of the assigned frequency.

§ 73.60 Frequency measurements.

(a) The carrier frequency of the transmitter shall be measured as often as necessary to ensure that it is maintained within the prescribed tolerance. However, in any event, the measurement shall be made at least once each calendar month with not more than 40 days expiring between successive measurements.

(b) The primary standard of frequency for radio frequency measurements shall be the national standard of frequency maintained by the National Bureau of Standards, Department of Commerce, Washington, D.C. The operating frequency of all radio stations will be determined by comparison with this standard or the standard signals of stations WWV, WWVB, WWVH and WWVL of the National Bureau of Standards.

§ 73.61 New equipment; restrictions.

The Commission will authorize the installation of new transmitting equipment in a broadcast station or changes in the frequency control of an existing transmitter only if such equipment is so designed that there is reasonable assurance that the transmitter is capable of maintaining automatically the assigned frequency within the limits specified in § 73.59.

§ 73.62 Automatic frequency control equipment; authorization required.

New automatic frequency control equipment and changes in existing automatic frequency control equipment that may affect the precision of frequency control or the operation of the transmitter shall be installed only upon authorization from the Commission.

§ 73.63 Auxiliary transmitter.

(a) Upon approval of an application therefor (Form 301), an auxiliary transmitter may be licensed for use either at the same location as the main transmitter or at another location, subject to the following conditions:

(1) A licensed operator shall be in control whenever an auxiliary transmitter is placed in operation.

(2) The auxiliary transmitter shall be maintained so that it may be placed in operation at any time for any one of the following purposes:

(i) The transmission of the regular programs upon the failure of the main transmitter.

(ii) The transmission of the regular programs during maintenance or modification work on the main transmitter necessitating discontinuance of its operation.

(iii) Emergency Broadcast System operation, provided the auxiliary transmitter is used in connection with an Emergency Broadcast System Authorization.

(iv) Upon request of a duly authorized representative of the Commission.

(v) An auxiliary transmitter may be used for the regular transmission of programs during periods of operation included in a Presunrise Service Authority (PSA).

(3) The auxiliary transmitter may be used only if it is in proper operating condition and adjusted to the licensed frequency. If testing is necessary to assure proper operation, it may be done any time. A dummy load or any authorized antenna may be used. Notations as to the time and results of such testing must be made in the maintenance log.

(4) The auxiliary transmitter shall be equipped with satisfactory control equipment which will enable the maintenance of the frequency emitted by the station within the limits prescribed by the regulations in this subpart.

(5) An auxiliary transmitter which is licensed at a geographical location different from that of the main transmitter shall be equipped with a frequency control which will automatically hold the frequency within the limits prescribed by the regulation in this part without any manual adjustment during operation or when it is being put into operation.

(6) The carrier frequency of the auxiliary transmitter shall be measured as often as is necessary to ensure that it is maintained within the prescribed tolerance. If the transmitter is used daily for a period of more than 40 days, the measurement shall be made at least once each calendar month with not more than 40 days expiring between successive measurements.

(7) The authorized antenna input power of an auxiliary transmitter may be less, but not more, than that of the regular transmitter. If it is less, the actual operating power is not limited to 105 percent of the authorized antenna input power of the auxiliary transmitter but shall in no event exceed the authorized antenna input power produced by the regular transmitter.

(8) All regulations as to safety requirements and spurious emissions applying to broadcast transmitting equipment shall apply also to an auxiliary transmitter.

(b) A licensee may, without further authority, utilize a replaced main transmitter as the auxiliary transmitter provided it remains installed at the same location as the main transmitter, and is coupled to the main antenna system and no changes in technical characteristics are involved. The Commission and the Engineer in Charge of the radio district in which the station is located must be notified within three days after it is ready for use as an auxiliary transmitter.

* * * * * * *

§ 73.67 Remote control operation.

(a) Operation by remote control shall be subject to the following conditions:

(1) The equipment at the operating and transmitting positions shall be so installed and protected that it is not accessible to or capable of operation by persons other than those duly authorized by the licensee.

(2) The control circuits from the operating positions to the transmitter shall provide positive on and off control and shall be such that open circuits, short circuits, grounds or other line faults will not actuate the transmitter and any fault causing loss of such control will automatically place the transmitter in an inoperative position.

(3) A malfunction of any part of the remote control system resulting in improper control shall be cause for the immediate cessation of operation by remote control. A malfunction of any part of the remote control system resulting in inaccurate meter readings, shall be cause for terminating operation by remote control no longer than 1 hour after the malfunction is detected.

(4) Control and monitoring equipment shall be installed so as to allow the licensed operator at the remote control point to perform all the functions in a manner required by the Commission's rules.

(5) Calibration of required indicating instruments at each remote control point shall be made against the corresponding instruments at the transmitter site at least once each calendar week. Results of calibrations shall be entered in the maintenance log. Remote control point meters shall be calibrated to provide an indication within 2 percent of the corresponding instrument reading at the transmitter site. In no event shall a remote control meter be calibrated against another remote meter.

(6) All remote control meters shall conform with specifications prescribed for regular transmitter, antenna, and monitor meters.

(7) Meters with arbitrary scale divisions may be used provided that calibration charts or curves are provided at the transmitter remote control point showing the relationship between the arbitrary scales and the reading of the main meters.

(b) All stations, whether operating by remote control or direct control, shall be equipped so as to be able to follow the Emergency Action Notification procedures described in § 73.911.

(c) The broadcast transmitter carrier may be amplitude modulated with a tone for the purpose of transmitting to the remote control point essential meter indications and other data on the operational condition of the broadcast transmitter and associated devices, subject to the following conditions:

(1) The tone shall have a frequency no higher than 30 hertz per second.

(2) The amplitude of modulation of the carrier by the tone shall not be higher than necessary to effect reliable and accurate data transmission, and shall not, in any case, exceed 6 percent.

(3) The tone shall be transmitted only at such times and during such intervals that the transmitted information is actually being observed or logged.

(4) Measures shall be employed to insure that during the periods the tone is being transmitted the total modulation of the carrier does not exceed 100 percent on negative peaks.

(5) Such tone transmissions shall not significantly degrade the quality of program transmission or produce audible effects resulting in public annoyance.

(6) Such tone transmissions shall not result in emissions of such a nature as to result in greater interference to other stations than is produced by normal program modulation.

§ 73.69 Antenna (phase) monitors.

(a) Each station utilizing a directional antenna shall have in operation at the transmitter an antenna monitor which is of a type approved by the Commission: *Provided, however,* That if the instrument of authorization of the station sets specific tolerances within which phase and amplitude relationships must be maintained, or requires the use of a monitor of specified repeatability or accuracy, the antenna monitor employed under such circumstances shall be authorized on an individual basis.

(b) In the event an antenna monitor becomes defective, the station may be operated without the monitor pending its repair or replacement for a period not in excess of 60 days without further authority from the Commission: *Provided*, That:

(1) Appropriate entries shall be made in the maintenance log of the station showing the date and time the monitor was removed and restored to service.

(2) If the license specifies antenna monitor sample current ratios, during the period the antenna monitor is out of service, base currents or remote base currents shall be read and logged at least once each day.

(3) Field strength measurements at each monitoring point specified in the station's authorization shall be read and logged at least once every 7 days.

(4) If the station is operated by remote control and phase indications are read and logged at the remote control point, during the period the antenna monitor is out of service indicating instruments at the transmitter shall be read and logged at the times specified in § 73.114(a)(9)(ii) for remotely controlled stations which do not provide for the reading and logging of phases at the remote control point.

(c) If conditions beyond the control of the licensee prevent the restoration of the monitor to service within the allowed period, informal request in accordance with § 1.549 of the Commission's rules must be filed with the Engineer in Charge of the radio district in which the station is located for such additional time as may be required to complete repairs of the defective instrument.

(d) If an authorized antenna monitor is replaced by another antenna monitor, the following procedure shall be followed:

(1) Temporary authority shall be requested and obtained from the Commission in Washington to operate with parameters at variance with licensed values, pending issuance of a modified license specifying new parameters.

(2) Immediately prior to the replacement of the antenna monitor, after a verification that all monitoring

point values and base current ratios are within the limits or tolerances specified in the instrument of authorization or the pertinent rules, the following indications shall be read and recorded in the maintenance log for each radiation pattern: Final plate current and plate voltage, common point current, base currents, antenna monitor phase and current indications, and the field strength at each monitoring point.

(3) With the new monitor substituted for the old, all indications specified in paragraph (c)(2) of this section, again shall be read and recorded. If no change has occurred in the indication for any parameter other than the indications of the antenna monitor the new antenna monitor indications shall be deemed to be those reflecting correct array adjustments.

(4) If it cannot be established by the observations required in paragraph (c)(2) of this section, that base current ratios and monitoring point values are within the tolerances or limits prescribed by the rules and the instrument of authorization, or if the substitution of the new antenna monitor for the old results in changes in these parameters, a partial proof of performance shall be executed, consisting of at least 10 field strength measurements, on each of the radials established in the latest complete adjustment of the antenna system. These measurements shall be made at locations, all within 2 to 10 miles from the antenna, which were utilized in such adjustment, including, on each radial, the location, if any, designated as a monitoring point in the station authorization. Measurements shall be analyzed in the manner prescribed in § 73.186.

(5) An informal request for modification of license shall be submitted to the Commission in Washington, D.C., within 30 days of the date of monitor replacement. Such request shall specify the make, type, and serial number of the replacement monitor, phase and sample current indications, and other data obtained pursuant to this paragraph (c).

NOTE: § 73.69(a) shall become effective, as follows:

(1) Each new station and each existing station for which major changes (see § 1.571(a)(1)) are authorized after June 1, 1973, shall be equipped with a type approved antenna monitor.

(2) Each station electing to utilize licensed operators other than first-class radiotelephone operators for routine transmitter duty (see § 73.93) shall meet this requirement by June 1, 1974. (Supply of type approved antenna monitors has been limited to the extent that not all licensees have been able to obtain delivery and install monitors by the June 1, 1974 deadline. Licensees deciding to use lower grade operators subsequent to June 1, 1974, have likewise been unable to obtain delivery. Therefore, such licensees will not be held accountable for failure to install the antenna monitor on evidence that timely efforts have been made to procure a monitor, and failure is due to non-delivery of equipment by suppliers. Each such licensee shall file a copy of its confirmed order with the Commission in Washington and retain a copy in the state file to be made available for inspection by FCC field engineers.)

(3) Each station operating by remote control, when adopting the schedule specified in § 73.114(a)(9)(iii) for observations at the transmitter, shall install a type-approved antenna monitor and provide phase indications at the remote control point, for observation and logging pursuant to § 73.113 (a)(3)(ii): *Provided,* That, in lieu of a type-approved monitor, the station may, until June 1, 1976, employ any monitor, manufactured after January 1, 1965, which is designed to afford phase indications on a device located external to the monitor, and which incorporates any necessary facilities whereby alternative RF inputs to the monitor may be selected by external switching.

(4) All other stations shall meet the requirements of this rule by June 1, 1977.

(e) The antenna monitor shall be calibrated once each calendar week according to manufacturer's instructions and a notation entered in the maintenance log.

* * * * * *

§ 73.71 Minimum operation schedule.

(a) All standard broadcast stations are required to maintain an operating schedule of not less than two-thirds of the total hours they are authorized to operate between 6 a.m. and 6 p.m., local time, and two-thirds of the total hours they are authorized to operate between 6 p.m. and midnight, local time, each day of the week except Sunday: *Provided, however,* That stations authorized for daytime operation only need comply only with the minimum requirement for operation between 6 a.m. and 6 p.m.

(b) In the event that causes beyond the control of a permittee or licensee make it impossible to adhere to the operating schedule in paragraph (a) of this section or to continue operating, the station may limit or discontinue operation for a period of not more than 30 days without further authority from the Commission, provided that notification is sent to the Commission in Washington, D.C. no later than the 10th day of limited or discontinued operation. During such period, the permittee or licensee shall continue to adhere to the requirements of the station license pertaining to the lighting of antenna structures. In the event normal operation is restored prior to the expiration of the 30 day period, the permittee or licensee will so notify the Commission in Washington, D.C. of this date. If causes beyond the control of the permittee or licensee make it impossible to comply within the allowed period, informal written request shall be made to the Commission in Washington, D.C., no later than the 30th day for such additional time as may be deemed necessary.

* * * * * *

§ 73.79 License to specify sunrise and sunset hours.

If the licensee of a broadcast station is required to commence or cease operation, or to change the mode of operation of the station, at the times of sunrise and sunset at any particular location, the controlling times for each month of the year are set forth in the station's instrument of authorization. Uniform sunrise and sunset times are specified for all of the days of each month, based upon the actual times of sunrise and sunset for the fifteenth day of the month adjusted to the nearest quarter hour. In accordance with a standardized procedure described therein, actual sunrise and sunset times are derived by interpolation in the tables of the 1946 American Nautical Almanac, issued by the Nautical Almanac Office of the United States Naval Observatory.

* * * * * * *

§ 73.92 Station and operator licenses; posting of.

(a) The station license and any other instrument of station authorization shall be posted in a conspicuous place and in such manner that all terms are visible, at the place the licensee considers to be the principal control point of the transmitter. At all other control points listed on the station authorization, a photocopy of the station license and other instruments of station authorization shall be posted.

(b) The operator license, or Form 759 (Verification of Operator License or Permit), of each station operator employed full-time, part-time or via contract, shall

167

be permanently posted and shall remain posted so long as the operator is employed by the licensee.

(1) The operator licenses shall be posted:

(i) Either at the transmitter or extension meter location; or

(ii) At the principal remote control point, if the station license authorizes operation by remote control.

(2) Posting of operator licenses shall be accomplished by affixing the license to the wall at the posting location, or enclosing in a binder, or inserting in folder and retaining at the posting location so that the licenses will be readily available and easily accessible at that location.

§ 73.93 Operator requirements.

(a) One or more operators holding a radio operator license or permit of a grade specified in this section shall be in actual charge of the transmitting system, and shall be on duty at the transmitter location, or at an authorized remote control point, or the position at which extension meters, as authorized pursuant to § 73.70 of this Subpart are located. The transmitter and required monitors and metering equipment, or the required extension meters and monitoring equipment and other required metering equipment, or the controls and required monitoring and metering equipment in an authorized remote control operation, shall be readily accessible to the licensed operator and located sufficiently close to the normal operating location that deviations from normal indications of required instruments can be observed from that location.

(b) With the exceptions set forth in paragraph (f) of this section, adjustments of the transmitting system, and inspection, maintenance, required equipment performance measurements, and required field strength measurements shall be performed only by a first-class radiotelephone operator, or, during periods of operation when a first-class radiotelephone operator is in charge of the transmitter, by or under the direction of a broadcast consultant regularly engaged in the practice of broadcast station engineering.

(c) A station using a non-directional antenna with nominal power of 10 kilowatts or less may employ first-class radiotelegraph operators, second-class radiotelegraph or radiotelephone operators, or operators with third-class radiotelegraph or radiotelephone permits endorsed for broadcast station operation for routine operation of the transmitting system if the station has at least one first-class radiotelephone operator readily available at all times. This operator may be in full-time employment or, as an alternative, the licensee may contract in writing for the services, on a part-time basis, of one or more such operators. Signed contracts with part-time operators shall be kept in the files of the station and shall be made available for inspection upon request by an authorized representative of the Commission.

(d) A station using a nondirectional antenna during periods of operation with authorized power in excess of 10 kilowatts, may employ first-class radiotelegraph operators, second-class radiotelegraph or radiotelephone operators, or radiotelegraph or radiotelephone operators with third-class permits endorsed for broadcast station operation, for routine operation of the transmitting system, if the station has in full-time employment at least one first-class radiotelephone operator.

(e) A station using a directional antenna system, which is required by the station authorization to maintain the ratio of the currents in the elements of the system within a tolerance which is less than 5 percent or the relative phases of those currents within a tolerance which is less than 3 degrees shall, without exemption, employ first-class radiotelephone operators who shall be on duty and in actual charge of the transmitting system as specified in paragraph (a) of this section during hours of operation with a directional radiation pattern. A station whose authorization does not specifically require therein the maintenance of phase and current relationships within closer tolerances than above specified shall employ first-class radiotelephone operators for routine operation of the transmitting system during periods of directional operation: *Provided, however*, That holders of first-class radiotelegraph licenses, second-class radiotelegraph or radiotelephone licenses, or third-class radiotelegraph or radiotelephone permits endorsed for broadcast station operation, may be employed for routine operation of the transmitting system, if the following conditions are met:

(1) The station must have in full-time employment at least one first-class radiotelephone operator.

(2) The station shall be equipped with a type-approved phase (antenna) monitor fed by a sampling system installed and maintained pursuant to accepted standards of good engineering practice (see Note 1 at end of this section).

(3) Within 1 year of the date on which lesser grade operators are first employed and a chief operator has been designated, pursuant to paragraph (h) of this section, the station shall complete a partial proof of performance, as defined in Note 2 at the end of this section, and shall complete subsequent partial proofs of performance at 3 year intervals thereafter. A skeleton proof of performance, as defined in Note 3 at the end of this section, shall be completed during each year that a partial proof of performance is not required. Not less than 10, nor more than 14 months shall elapse between the completion dates of successive proofs of performance. The results of such proofs shall be prepared and filed in the same manner as required in equipment performance measurements pursuant to paragraph (b) of § 73.47.

NOTE 1: See § 73.69, NOTE: (2).

(4) Field strength measurements shall be made at the monitoring points specified in the station authorization at least once each 30 days unless more frequent measurements are required by such authorization. The results of these measurements shall be entered in the station maintenance log. The licensee shall have readily available, and in proper working condition, field strength measuring equipment to perform these measurements.

(f) Subject to the conditions set forth in paragraphs (c), (d) and (e) of this section, the routine operation of the transmitting system may be performed by an operator holding a first-class radiotelegraph license, a second-class radiotelegraph or radiotelephone license, or a third-class radiotelegraph or radiotelephone permit endorsed for broadcast operation. Unless, however, performed under the immediate and personal supervision of an operator holding a first-class radiotelephone license, an operator holding a first-class radiotelegraph license, second-class radiotelegraph or radiotelephone license, or third-class radiotelegraph or radiotelephone permit endorsed for broadcast station operation, may make adjustments only of external controls, as follows:

(1) Those necessary to turn the transmitter on and off;

(2) Those necessary to compensate for voltage fluctuations in the primary power supply;

(3) Those necessary to maintain modulations levels of the transmitter within prescribed limits;

(4) Those necessary to effect routine changes in operating power which are required by the station authorization;

(5) Those necessary to change between nondirectional and directional or between differing radiation patterns, provided that such changes require only activation of switches and do not involve the manual tuning of the transmitter final amplifier or antenna phasor equipment. The switching equipment shall be so arranged that the failure of any relay in the directional antenna system to activate properly will cause the emissions of the station to terminate.

(g) It is the responsibility of the station licensee to insure that each operator is fully instructed in the performance of all the above adjustments, as well as in other required duties, such as reading meters and making log entries. Printed step-by-step instructions for those adjustments which the lesser grade operator is permitted to make, and a tabulation or chart of upper and lower limiting values of parameters required to be observed and logged, shall be posted at the operating position. The emissions of the station shall be terminated immediately whenever the transmitting system is observed operating beyond the posted parameters, or in any other manner inconsistent with the rules or the station authorization, and the above adjustments are ineffective in correcting the condition of improper operation, and a first-class radiotelephone operator is not present.

(h) When lesser grade operators are used, in accordance with paragraph (d) or (e) of this section, for any period of operation with nominal power in excess of 10 kilowatts, or with a directional radiation pattern, the station licensee shall designate one first-class radiotelephone operator in full-time employment as the chief operator who, together with the licensee, shall be responsible for the technical operation of the station. The licensee also may designate another first-class radiotelephone operator as assistant chief operator, who shall assume all responsibilities of the chief operator during periods of his absence. The station licensee shall notify the engineer in charge of the radio district in which the station is located of the name(s) and license number(s) of the operator(s) so designated. Such notification shall be made within 3 days of the date of such designation. A copy of the notification shall be posted with the license(s) of the designated operator(s).

(1) An operator designated as chief operator for one station may not be so designated concurrently at any other standard broadcast station.

(2) The station licensee shall vest such authority in, and afford such facilities to the chief operator as may be necessary to insure that the chief operator's primary responsibility for the proper technical operation of the station may be discharged efficiently.

(3) At such time as the regularly designated chief operator is unavailable or unable to act as chief operator (e.g., vacations, sickness), and an assistant chief operator has not been designated, or, if designated, for any reason is unable to assume the duties of the chief operator, the licensee shall designate another first-class radiotelephone operator as acting chief operator on a temporary basis. Within 3 days of the date such action is taken, the engineer in charge of the radio district in which the station is located shall be notified by the licensee by letter of the name and license number of the acting chief operator, and shall be notified by letter, again within 3 days of the date when the regularly designated chief operator returns to duty.

(4) The designated chief operator may serve as a routine duty transmitter operator at any station only to the extent that it does not interfere with the efficient discharge of his responsibilities as listed below.

(i) The inspection and maintenance of the transmitting system including the antenna system and required monitoring equipment.

(ii) The accuracy and completeness of entries in the maintenance log.

(iii) The supervision and instruction of all other station operators in the performance of their technical duties.

(iv) A review of completed operating logs to determine whether technical operation of the station has been in accordance with the rules and terms of the station authorization. After review, the chief operator shall sign the log and indicate the date and time of such review. If the review of the operating logs indicates technical operation of the station is in violation of the rules or terms of the station authorization, he shall promptly initiate corrective action. The review of each day's operating log shall be made within 24 hours, except that, if the chief operator is not on duty during a given 24 hour period, the logs must be reviewed within 2 hours after his next appearance for duty. In any case, the time before review shall not exceed 72 hours.

(i) The operator on duty at the transmitter or remote control point, may, at the discretion of the licensee and the chief operator, if any, be employed for other duties or for the operation of another radio station or stations in accordance with the class of operator's license which he holds and the rules and regulations governing such other stations: *Provided, however*, That such other duties shall not interfere with the proper operation of the standard broadcast transmitting system and keeping of required logs.

(j) At all standard broadcast stations, a complete inspection of the transmitting system and required monitoring equipment in use, shall be made by an operator holding a first-class radiotelephone license at least once each calendar week. The interval between successive required inspections shall not be less than 5 days. This inspection shall include such tests, adjustments, and repairs as may be necessary to insure operation in conformance with the provisions of this subpart and the current station authorization.

NOTE 1: See § 73.69.

NOTE 2: The partial proof of performance shall consist of at least 10 field strength measurements on each of the radials established in the latest complete adjustment of the directional antenna system. These measurements shall be made at locations, all within 2 to 10 miles from the antenna, which were utilized in such adjustments, and include on each radial, the point, if any, designated as a monitoring point in the station authorization. Measurements shall be analyzed in the manner prescribed in § 73.186 of the rules of this part.

NOTE 3: The skeleton proof of performance shall consist of field strength measurements, at least three on each of the radials established in the latest complete adjustment of the directional antenna system, made at measurement locations utilized in such adjustment, and include, on each radial, the point, if any, designated as a monitoring point in the station authorization.

§ 73.95   Equipment tests.

(a) During the process of construction of a standard broadcast station the permittee, after notifying the Commission and Engineer in Charge of the radio district in which the station is located, may, without further authority of the Commission, conduct equipment tests during the experimental period for the purpose of such adjustments and measurements as may be necessary to assure compliance with the terms of the construction permit, the technical provisions of the application therefor, the rules and regulations, and the applicable engineering standards. In addition the Commission may authorize equipment tests other than during the experimental period if such operation is shown to be desirable to the proper completion of construction and adjustment of the transmitting equipment and antenna system. An informal application for such authority, giving full details regarding the need for such tests, shall be filed with the Commission at least two (2) days (not including Sundays and Saturdays and legal holidays when the offices of the Commission are not open) prior to the date on which it is desired to begin such operation.

(b) The Commission may notify the permittee to conduct no tests or may cancel, suspend, or change the date for the beginning of equipment tests as and when such action may appear to be in the public interest, convenience, or necessity.

(c) Equipment tests may be continued so long as the construction permit shall remain valid and shall be conducted only during the experimental period (12 midnight to local sunrise) unless otherwise specifically authorized.

(d) Inspection of a station will ordinarily be required during the equipment test period and before the commencement of program tests. After construction and after adjustments and measurements have been completed to show compliance with the terms of the construction permit, the technical provisions of the application therefor, the rules and regulations and the applicable engineering standards, the permittee should notify the Engineer in Charge of the radio district in which the station is located that it is ready for inspection.

(e) The authorization for tests embodied in this section shall not be construed as constituting a license to operate but as a necessary part of construction.

*   *   *   *   *   *   *

§ 73.97   Station inspection.

The licensee of any radio station shall make the station available for inspection by representatives of the Commission at any reasonable hour.

§ 73.98   Operation during emergency.

(a) When necessary to the safety of life and property and in response to dangerous conditions of a general nature, standard broadcast stations may, at the discretion of the licensee and without further Commission authority, transmit emergency weather warnings and other emergency information. Examples of emergency situations which may warrant either an immediate or delayed response by the licensee are: Tornadoes, hurricanes, floods, tidal waves, earthquakes, icing conditions, heavy snows, widespread fires, discharge of toxic gases, widespread power failures, industrial explosions, and civil disorders. Transmission of information concerning school closings and changes in schoolbus schedules resulting from any of these conditions, is appropriate. In addition, and if requested by responsible public officials, emergency point-to-point messages may be transmitted for the purpose of requesting or dispatching aid and assisting in rescue operations.

(b) When emergency operation is conducted under a State-Level EBS Operational Plan, the attention signal described in § 73.906 may be employed.

(c) Except as provided in paragraph (d) of this section, emergency operation shall be confined to the hours, frequencies, powers, and modes of operation specified in the license documents of the stations concerned.

(d) When adequate advance warning cannot be given with the facilities or hours authorized, stations may employ their full daytime facilities during nighttime hours to carry weather warnings and other types of emergency information connected with the examples listed in paragraph (a) of this section. Because of skywave interference impact on other stations assigned to the same channel, such operation may be undertaken only if regular, unlimited-time service is nonexistent, inadequate from the standpoint of coverage, or not serving public need. All operation under this paragraph must be conducted on a noncommercial basis. Recorded music may be used to the extent necessary to provide program continuity.

(e) Any emergency operation undertaken in accordance with this section may be terminated by the Commission, if required in the public interest.

(f) Immediately upon cessation of an emergency during which broadcast facilities were used for the transmission of point-to-point messages under paragraph (a) of this section, or when daytime facilities were used during nighttime hours in accordance with paragraph (d) of this section, a report in letter form shall be forwarded to the Commission, in Washington, D.C., setting forth the nature of the emergency, the dates and hours of emergency operation, and a brief description of the material carried during the emergency period. A certification of compliance with the noncommercialization provision of paragraph (d) of this section must accompany the report where daytime facilities are used during nighttime hours, together with a detailed showing concerning the alternate service provisions of that paragraph.

(g) If the Emergency Broadcast System (EBS) is activated at the National-Level while non-EBS emergency operation under this section is in progress, the EBS shall take precedence.

*   *   *   *   *   *   *

§ 73.111   General requirements relating to logs.

(a) The licensee or permittee of each standard broadcast station shall maintain program, operating, and maintenance logs as set forth in §§ 73.112, 73.113, and 73.114. Each log shall be kept by the station employee or employees (or contract operator) competent to do so, having actual knowledge of the facts required, who in the case of program and operating logs shall sign the appropriate log when starting duty, and again when going off duty.

(b) The logs shall be kept in an orderly and legible manner, in suitable form, and in such detail that the data required for the particular class of station concerned is readily available. Key letters or abbreviations may be used if proper meaning or explanation is contained elsewhere in the log. Each sheet shall be numbered and dated. Time entries shall be made in local time. For the period from the last Sunday in April until the last Sunday in October of each year, the program and operating log entries showing times of sign-on, sign-off, and change in the station's mode of operation shall specifically be indicated as advanced or nonadvanced time.

(c) No log or preprinted log or schedule which becomes a log, or portion thereof, shall be erased, obliterated, or willfully destroyed within the period of retention provided by the provisions of this part. Any necessary correction shall be made only pursuant to §§ 73.112, 73.113, and 73.114, and only by striking out the erroneous portion, or by making a corrective explanation on the log or attachment to it as provided in those sections.

(d) Entries shall be made in the logs as required by §§ 73.112, 73.113, and 73.114. Additional information such as that needed for billing purposes or for the cuing of automatic equipment may be entered on the logs. Such additional information, so entered, shall not be subject to the restrictions and limitations in the Commission's rules on the making of corrections and changes in logs.

(e) The operating log and the maintenance log may be kept individually on the same sheet in one common log, at the option of the permittee or licensee.

§ 73.112 Program log.

(a) The following entries shall be made in the program log:

(1) *For each program.* (i) An entry identifying the program by name or title.

(ii) An entry of the time each program begins and ends. If programs are broadcast during which separately identifiable program units of a different type or source are presented, and if the licensee wishes to count such units separately, the beginning and ending time for the longer program need be entered only once for the entire program. The program units which the licensee wishes to count separately shall then be entered underneath the entry for a longer program, with the beginning and ending time of each such unit, and with the entry indented or otherwise distinguished so as to make it clear that the program unit referred to was broadcast within the longer program.

(iii) An entry classifying each program as to type, using the definitions set forth in Note 1 at the end of this section.

(iv) An entry classifying each program as to source, using the definitions set forth in Note 2 at the end of this section. (For network programs, also give name or initials of network, e.g., ABC, CBS, NBC, Mutual.)

(v) An entry for each program presenting a political candidate, showing the name and political affiliation of such candidate.

(2) *For commercial matter.* (i) An entry identifying (a) the sponsor(s) of the program, (b) the person(s) who paid for the announcement, or (c) the person(s) who furnished materials or services referred to in § 73.119(d). If the title of a sponsored program includes the name of the sponsor, e.g., XYZ News, a separate entry for the sponsor is not required.

(ii) An entry or entries showing the total duration of commercial matter in each hourly time segment (beginning on the hour) or the duration of each commercial message (commercial continuity in sponsored programs, or commercial announcements) in each hour. See Note 5 at the end of this section for statement as to computation of commercial time.

(iii) An entry showing that the appropriate announcement(s) (sponsorship, furnishing material or services, etc.) have been made as required by section 317 of the Communications Act and § 73.119. A checkmark (✓) will suffice but shall be made in such a way as to indicate the matter to which it relates.

(3) *For public service announcements.* (i) An entry showing that a public service announcement (PSA) has been broadcast together with the name of the organization or interest on whose behalf it is made. See Note 4 following this section for definition of a public service announcement.

(4) *For other announcements.* (i) an entry of the time that each required station identification announcement is made (call letters and licensed location; see § 73.1201).

(ii) An entry for each announcement presenting a political candidate showing the name and political affiliation of such candidate.

(iii) An entry for each announcement made pursuant to the local notice requirements of §§ 1.580 (pregrant) and 1.594 (designation for hearing) of this chapter, showing the time it was broadcast.

(iv) An entry showing that broadcast of taped, filmed, or recorded material has been made in accordance with the provisions of § 73.1208.

(b) Program log entries may be made either at the time of or prior to broadcast. A station broadcasting the programs of a national network which will supply it with all information as to such programs, commercial matter and other announcements for the composite week need not log such data but shall record in its log the time when it joined the network, the name of each network program broadcast, the time it leaves the network, and any nonnetwork matter broadcast required to be logged. The information supplied by the network, for the composite week which the station will use in its renewal application, shall be retained with the program logs and associated with the log pages to which it relates.

(c) No provision of this section shall be construed as prohibiting the recording or other automatic maintenance of data required for program logs. However, where such automatic logging is used, the licensee must comply with the following requirements:

(1) The licensee, whether employing manual or automatic logging or a combination thereof, must be able accurately to furnish the Commission with all information required to be logged;

(2) Each recording, tape, or other means employed shall be accompanied by a certificate of the operator or other responsible person on duty at the time or other duly authorized agent of the licensee, to the effect that it accurately reflects what was actually broadcast. Any information required to be logged which cannot be incorporated in the automatic process shall be maintained in a separate record which shall be similarly authenticated;

(3) The licensee shall extract any required information from the recording for the days specified by the Commission or its duly authorized representative and submit it in written log form, together with the underlying recording, tape, or other means employed.

(d) Program logs shall be changed or corrected only in the manner prescribed in § 73.111(c) and only in accordance with the following:

(1) *Manually kept log.* Where, in any program log, or preprinted program log, or program schedule which upon completion is used as a program log, a correction is made before the person keeping the log has signed the log upon going off duty, such correction, no matter by whom made, shall be initialed by the person keeping the log prior to his signing of the log when going off duty, as attesting to the fact that the log as corrected is an accurate representation of what was broadcast. If corrections or additions are made on the log after it has been so signed, explanation must be made on the log or on an attachment to it, dated and signed by either the person who kept the log, the station program director or manager, or an officer of the licensee.

NOTE 1. *Program type definitions.* The definitions of the first eight types of programs (a) through (h) are intended not to overlap each other and will normally include all the various programs broadcast. Definitions (i) through (k) are subcategories and the programs classified thereunder will also be classified under one of the appropriate first eight types. There may also be further duplication within types (i) through (k); (e.g., a program presenting a candidate for public office, prepared by an educational institution, would be classified as Public Affairs (PA), Political (POL), and Educational Institutions (ED)).

(a) Agricultural programs (A) include market reports, farming, or other information specifically addressed, or primarily of interest, to the agricultural population.

(b) Entertainment programs (E) include all programs intended primarily as entertainment, such as music, drama, variety, comedy, quiz, etc.

(c) News programs (N) include reports dealing with current local, national, and international events, including weather and stock market reports; and when an integral part of a news program, commentary, analysis, and sports news.

(d) Public affairs programs (PA) include talks, commentaries, discussions, speeches, editorials, political programs, documentaries, forums, panels, roundtables, and similar programs primarily concerning local, national, and international public affairs.

(e) Religious programs (R) include sermons or devotionals; religious news; and music, drama, and other types of programs designed primarily for religious purposes.

(f) Instructional programs (I) include programs (other than those classified under Agricultural, News, Public Affairs, Religious or Sports) involving the discussion of, or primarily designed to further an appreciation or understanding of, literature, music, fine arts, history, geography, and the natural and social sciences; and programs devoted to occupational and vocational instruction, instruction with respect to hobbies, and similar programs intended primarily to instruct.

(g) Sports programs (S) include play-by-play and pre- or post-game related activities and separate programs of sports instruction, news or information (e.g., fishing opportunities, golfing instruction, etc.).

(h) Other programs (O) include all programs not falling within definitions (a) through (g).

(i) Editorials (EDIT) include programs presented for the purpose of stating opinions of the licensee.

(j) Political programs (POL) include those which present candidates for public office or which give expressions (other than in station editorials) to views on such candidates or on issues subject to public ballot.

(k) Educational Institution programs (ED) include any program prepared by, in behalf of, or in cooperation with, educational institutions, educational organizations, libraries, museums, PTA's, or similar organizations. Sports programs shall not be included.

NOTE 2. *Program source definitions.*—(a) A local program (L) is any program originated or produced by the station, or for the production of which the station is primarily responsible, employing live talent more than 50 percent of the time. Such a program, taped or recorded for later broadcast, shall be classified as local. A local program fed to a network shall be classified by the originating station as local. All nonnetwork news programs may be classified as local. Programs primarily featuring records or transcriptions shall be classified as recorded (REC) even though a station announcer appears in connection with such material. However, identifiable units of such programs which are live and separately logged as such may be classified as local. (E.g., if during the course of a program featuring records or transcriptions a nonnetwork 2-minute news report is given and logged as a news program, the report may be classified as local.)

(b) A network program (NET) is any program furnished to the station by a network (national, regional or special). Delayed broadcasts of programs originated by networks are classified as network.

(c) A recorded program (REC) is any program not otherwise defined in this Note including, without limitation, those using recordings, transcriptions or tapes.

NOTE 3. *Definition of commercial matter* (CM) includes commercial continuity (network and nonnetwork) and commercial announcements (network and nonnetwork) as follows: (Distinction between continuity and announcements is made only for definition purposes. There is no need to distinguish between the two types of commercial matters when logging.)

(a) Commercial continuity (CC) is the advertising message of a program sponsor.

(b) A commercial announcement (CA) is any other advertising message for which a charge is made, or other consideration is received.

(1) Included are (i) "bonus spots"; (ii) trade-out spots, and (iii) promotional announcements of a future program where consideration is received for such an announcement or where such announcement identifies the sponsor of a future program beyond mention of the sponsor's name as an integral part of the title of the program. (E.g., where the agreement for the sale of time provides that the sponsor will receive promotional announcements, or when the promotional announcement contains a statement such as "Listen tomorrow for the—[name of program]—brought to you by—[sponsor's name]—.")

(2) Other announcements, including but not limited to the following, are not commercial announcements:

(i) Promotional announcements, except as heretofore defined in paragraph (b).

(ii) Station identification announcements for which no charge is made.

(iii) Mechanical reproduction announcements.

(iv) Public service announcements.

(v) Announcements made pursuant to § 73.119(d) that materials or services have been furnished as an inducement to broadcast a political program or a program involving the discussion of controversial public issues.

(vi) Announcements made pursuant to the local notice requirements of §§ 1.580 (pre-grant) and 1.594 (designation for hearing) of this chapter.

NOTE 4. *Definition of a public service announcement.* A public service announcement is an announcement for which no charge is made and which promotes programs, activities, or services of Federal, State, or local governments (e.g., recruiting, sales of bonds, etc.) or the programs, activities or services of nonprofit organizations (e.g., UGF, Red Cross Blood Donations, etc.), and other announcements regarded as serving community interests, excluding time signals, routine weather announcements, and promotional announcements.

NOTE 5. *Computation of commercial time.* Duration of commercial matter shall be as close an approximation to the time consumed as possible. The amount of commercial time scheduled will usually be sufficient. It is not necessary, for example, to correct an entry of a 1-minute commercial to accommodate varying reading speeds even though the actual time consumed might be a few seconds more or less than the scheduled time. However, it is incumbent upon the licensee to ensure that the entry represents as close an approximation of the time actually consumed as possible.

§ 73.113 Operating log.

(a) Entries shall be made in the operating log either manually by a properly licensed operator in actual charge of the transmitting apparatus, or by automatic devices meeting the requirements of paragraph (b) of this section. Indications of operating parameters shall be logged prior to any adjustment of the equipment. Where adjustments are made to restore parameters to their proper operating values, the corrected indications shall be logged, accompanied, if any parameter deviation was beyond a prescribed tolerance, by a notation describing the nature of the corrective action. Indications of all parameters whose values are affected by modulation of the carrier shall be read without modulation. The actual time of observation shall be included in each log entry. The operating log shall include the following information:

(1) For all stations:

(i) Entries of the time the station begins to supply power to the antenna and the time it ceases to do so.

(ii) Entries required by § 17.49 (a), (b), and (c) of this chapter concerning daily observations of tower lights.

(iii) Any entries not specifically required in this section, but required by the instrument of authorization or elsewhere in this part. See, particularly, the additional entries required by § 73.51(e)(2) when power is being determined by the indirect method.

(iv) The following indications shall be entered in the operating log at the time of commencement of operation in each mode and thereafter, at successive intervals not exceeding 3 hours in duration.

(v) A notation of tests of the Emergency Broadcast System procedures pursuant to the requirements of Subpart G of this Part and the appropriate station EBS checklist.

(a) Total plate voltage and total plate current of the last radio stage.

(b) Antenna current or remote antenna current (for nondirectional operation); common point current or remote common point current (for directional operation).

(2) For stations with directional antennas not operated by remote control, the following indications, in addition to those specified in subparagraph (1) of this paragraph, shall be read and entered in the operating log at the time of commencement of operation in each mode and thereafter, at successive intervals not exceeding 3 hours in duration. (This schedule shall apply regardless of any provision in the station instrument of authorization requiring more frequent log entries.)

(i) Phase indications.

(ii) Remote antenna base current or antenna monitor sample current or current ratio indications.

(3) For stations with directional antennas operated by remote control, the following indications, in addition to those specified in subparagraph (1) of this paragraph, shall be read and entered in the operating log at the time of commencement of operation in each mode and thereafter, at successive intervals not exceeding 3 hours in duration. (This schedule shall apply, regardless of any provision in the station instrument of authorization requiring more frequent log entries.)

(i) Either remote indications of base currents, or currents extracted from antenna monitor sampling lines, or current indications or their ratios provided by a type-approved antenna monitor.

(ii) Phase indications, if provided by a type-approved antenna monitor.

(b) Automatic devices accurately calibrated and with appropriate time, date and circuit functions may be utilized to record the entries in the operating log: *Provided*, That:

(1) They do not affect the operation of circuits or accuracy of indicating instruments of the equipment being recorded;

(2) The recording devices have an accuracy equivalent to the accuracy of the indicating instruments;

(3) The calibration is checked against the original indicators at least once a week and the results noted in the maintenance log;

(4) Provision is made to actuate automatically an aural alarm circuit located near the operator on duty if any of the automatic log readings are not within the tolerances or other requirements specified in the rules or instrument of authorization;

(5) Unless the alarm circuit operates continuously, devices which record each parameter in sequence must read each parameter at least once during each 10-minute period and clearly indicate the parameter being recorded;

(6) The automatic logging equipment is located at the remote control point if the transmitter is remotely controlled, or at the transmitter location if the transmitter is directly controlled;

(7) The automatic logging equipment is located in the near vicinity of the operator on duty and is inspected by him periodically during the broadcast day; and

(8) The indicating equipment conforms with the requirements of § 73.39 except that the scales need not exceed 2 inches in length. Arbitrary scales may not be used.

(c) In preparing the operating log, original data may be recorded in rough form and later transcribed into the log.

(d) Operating logs shall be changed or corrected only in the manner prescribed in § 73.111(c) and only in accordance with the following:

(1) *Manually kept log.* Any necessary corrections in a manually kept operating log shall be made only by the person making the original entry who shall make and initial each correction prior to signing the log when going off duty in accordance with § 73.111(a). If corrections or additions are made on the log after it has been so signed, explanation must be made on the log or on an attachment to it, dated and signed by either the operator who kept the log, the station technical supervisor or an officer of the licensee.

(2) *Automatic logging.* No automatically kept operating log shall be altered in any way after entries have been recorded. Any errors or omissions found

in an automatically kept operating log shall be noted and explained in a memorandum signed by the operator on duty (who, under the provisions of paragraph (b)(7) of this section, is required to inspect the automatic equipment), or by the station technical supervisor or an officer of the licensee. Such memorandum shall be affixed to the original log in question.

(e) If required by § 73.93(h)(4)(iv), each completed operating log shall bear a signed notation by the station's chief operator of the results of the review of that log, and show the date and time of such review.

* * * * * *

§ 73.115 Retention of logs.

Logs of standard broadcast stations shall be retained by the licensee or permittee for a period of 2 years: *Provided, however,* That logs involving communications incident to a disaster or which include communications incident to or involved in an investigation by the Commission and concerning which the licensee or permittee has been notified, shall be retained by the licensee or permittee until he is specifically authorized in writing by the Commission to destroy them: *Provided, further,* That logs incident to or involved in any claim or complaint of which the licensee or permittee has notice shall be retained by the licensee or permittee until such claim or complaint has been fully satisfied or until the same has been barred by statute limiting the time for the filing of suits upon such claims.

NOTE: Application forms for licenses and other authorizations require that certain operating and program data be supplied. It is suggested that these application forms be kept in mind in connection with maintenance of station program and operating records.

§ 73.116 Availability of logs and records.

The following shall be made available upon request by an authorized representative of the Commission:

(a) Program, operating and maintenance logs.

(b) Equipment performance measurements required by § 73.47.

(c) Copy of the most recent antenna resistance or common-point impedance measurements submitted to the Commission.

(d) Copy of the most recent field intensity measurements to establish performance of directional antennas required by § 73.151.

* * * * * *

SUBPART B—FM BROADCAST STATIONS

* * * * * *

§ 73.242 Duplication of AM and FM programing.

(a) Licensees of FM stations in cities of over 100,000 population (as listed in the latest regular U.S. Census Reports) shall operate so as to devote no more than 50 percent of the average FM broadcast week to programs duplicated from an AM station owned by the same licensee in the same local area. For the purposes of this paragraph, duplication is defined to mean simultaneous broadcasting of a particular program over both the AM and FM station or the broadcast of a particular FM program within 24 hours before or after the identical program is broadcast over the AM station.

(b) Compliance with the non-duplication requirement shall be evidenced by such showing in connection with renewal applications as the Commission may require.

(c) Upon a substantial showing that continued program duplication over a particular station would better serve the public interest than immediate non-duplication, a licensee may be granted a temporary exemption from the requirements of paragraph (a) of this section. Requests for such exemption must be submitted to the Commission, accompanied by supporting data, at least 6 months prior to the time the non-duplication requirement of paragraph (a) of this section is to become effective as to a particular station. Such exemption, if granted, will ordinarily run to the end of the station's current license period, or if granted near the end of the license period, for some other reasonable period not to exceed 3 years.

* * * * * *

§ 73.253 Modulation monitors.

(a) Each station shall have in operation either at the transmitter or the extension meter location, or at the place where the transmitter is controlled, a modulation monitor of a type approved by the Commission for non-multiplex operation: *Provided,* That: (1) If the station is engaged in stereophonic operation as contemplated by § 73.297, the licensee shall have in operation a modulation monitor of a type approved by the Commission for monitoring stereophonic operation, and (2) if the station is engaged in operation with a Subsidiary Communications Authorization, as contemplated by § 73.295, the licensee shall have in operation a modulation monitor of a type approved by the Commission for monitoring SCA operation.

NOTE: Approved modulation monitors (non-multiplex, stereophonic, and SCA) are included on the Commission's "Radio Equipment List". Copies of this list are available for inspection at the Commission's offices in Washington, D.C. and at each of its field offices.

(b) In the event that the modulation monitor becomes defective, the station may be operated without the monitor pending its repair or replacement for a period not in excess of 60 days without further authority of the Commission: *Provided,* That:

(1) Appropriate entries shall be made in the maintenance log of the station showing the date and time the monitor was removed and restored to service.

(2) During the period when the station is operated without the modulation monitor, the licensee shall provide other suitable means for insuring that the modulation is maintained within the tolerance prescribed in § 73.268.

(c) If conditions beyond the control of the licensee prevent the restoration of the monitor to service within the above allowed period, informal request in accordance with § 1.549 of this chapter may be filed with the Engineer in Charge of the radio district in which the station is located for such additional time as may be required to complete repairs of the defective instrument.

* * * * * *

§ 73.258 Indicating instruments.

(a) Each FM broadcast station shall be equipped with indicating instruments, which conform with the specifications set forth in § 73.320, for measuring the direct plate voltage and current of the last radio stage and the transmission line radio frequency current, voltage or power.

(b) In the event that any one of these indicating instruments becomes defective when no substitute which conforms with the required specifications is available, the station may be operated without the defective instrument pending its repair or replacement for a period not in excess of 60 days: *Provided, That*:

(1) Appropriate entries shall be made in the maintenance log of the station showing the date and time the meter was removed from and restored to service.

(2) [Reserved]

(3) If the defective instrument is a plate voltmeter or plate ammeter in the last radio stage, the operating power shall be maintained by means of the radio frequency transmission line meter.

(c) If conditions beyond the control of the licensee prevent the restoration of the meter to service within the above allowed period, informal request may be filed in accordance with § 1.549 of this chapter with the Engineer in Charge of the radio district in which the station is located for such additional time as may be required to complete repairs of the defective instrument.

§ 73.261 Time of operation.

(a) All FM broadcast stations will be licensed for unlimited time operation. All FM stations are required to maintain an operating schedule of not less than 8 hours between 6 a.m. and 6 p.m., local time, and not less than 4 hours between 6 p.m. and midnight, local time, each day of the week except Sunday.

(b) In the event that causes beyond the control of a permittee or licensee make it impossible to adhere to the operating schedule in paragraph (a) of this section or to continue operating, the station may limit or discontinue operation for a period of not more than 30 days without further authority from the Commission, *Provided,* That notification is sent to the Commission in Washington, D.C. no later than the 10th day of limited or discontinued operation. During such period, the permittee or licensee shall continue to adhere to the requirements of the station license pertaining to the lighting of antenna structures. In the event normal operation is restored prior to the expiration of the 30 day period, the permittee or licensee will so notify the Commission in Washington, D.C. of this date. If the causes beyond the control of the permittee or licensee make it impossible to comply within the allowed period, informal written request shall be made to the Commission in Washington, D.C., no later than the 30th day for such additional time as may be deemed necessary.

* * * * * * *

§ 73.263 Station inspection.

The licensee of any FM broadcast station shall make the station available for inspection by representatives of the Commission at any reasonable hour.

§ 73.264 Station and operator licenses; posting of.

(a) The station license and any other instrument of station authorization shall be posted in a conspicuous place and in such manner that all terms are visible, at the place the licensee considers to be the principal control point of the transmitter. At all other control points listed on the station authorization, a photocopy of the station license and other instruments of station authorization shall be posted.

(b) The operator license, or Form 759 (Verification of Operator License or Permit), of each station operator employed full-time, part-time or via contract, shall be permanently posted and shall remain posted so long as the operator is employed by the licensee.

(1) The operator licenses shall be posted:

(i) Either at the transmitter or extension meter location; or

(ii) At the principal remote control point, if the station license authorizes operation by remote control.

(2) Posting of operator licenses shall be accomplished by affixing the license to the wall at the posting location, or enclosing in a binder, or inserting in folder and retaining at the posting location so that the licenses will be readily available and easily accessible at that location.

§ 73.265 Operator requirements.

(a) One or more operators holding a radio operator license or permit of a grade specified in this section shall be in actual charge of the transmitting system, and shall be on duty at the transmitter location, or at an authorized remote control point, or the position at which extension meters, as authorized pursuant to § 73.276 of this Subpart, are located. The transmitter and required monitors and metering equipment, or the required extension meters and monitoring equipment, or the controls and required monitoring and metering equipment in an authorized remote control operation, shall be readily accessible to the licensed operator and located sufficiently close to the normal operating location that deviations from normal indications of required instruments can be observed from that location.

(b) With the exceptions set forth in paragraph (e) of this section, adjustments to the transmitting system, and inspection, maintenance, and required equipment performance measurements shall be performd only by a first-class radiotelephone operator, or, during periods of operation when a first-class radiotelephone operator is in charge of the transmitter, by, or under the direction of a broadcast consultant who is regularly engaged in the practice of broadcast station engineering.

(c) A station with authorized transmitter output power of 25 kilowatts or less may employ first-class radiotelegraph operators, second-class radiotelegraph or radiotelephone operators, or operators with third-class radiotelegraph or radiotelephone permits endorsed for broadcast station operation for routine operation of the transmitting system if the station has at least one first-class radiotelephone operator readily available at all times. This operator may be in full-time employment or, as an alternative, the licensee may contract in writing for the services on a part-time basis, of one or more such operators. Signed contracts with part-time operators shall be kept in the files of the station and shall be made available for inspection upon request by an authorized representative of the Commission.

(d) A station with authorized transmitter output power in excess of 25 kilowatts may employ first-class radiotelegraph operators, second-class radiotelegraph or radiotelephone operators, or operators with third-class radiotelegraph or radiotelephone permits endorsed for broadcast station operation for routine operation of the transmitting system if the station has in full-time employment at least one first-class radiotelephone operator and complies with the following:

(1) The station licensee shall designate one first-class radiotelephone operator in full-time employment as the chief operator who, together with the licensee, shall be responsible for the technical operation of the station. The licensee may also designate another first-class radiotelephone operator as assistant chief operator, who shall assume all responsibilities of the chief operator during periods of his absence. The station licensee shall notify the engineer in charge of the radio district in which the station is located of the name(s) and license number(s) of the operator(s) so designated. Such notification shall be made within 3 days of the date of such designation. A copy of the notification shall be posted with the license(s) of the designated operator(s).

(2) An operator designated as chief operator for one station may not be so designated concurrently at any other FM broadcast station.

(3) The station licensee shall vest such authority in, and afford such facilities to the chief operator as may be necessary to insure that the chief operator's primary responsibility for the proper technical operation of the station may be discharged efficiently.

(4) At such times as the regularly designated chief operator is unavailable or unable to act as chief operator (e.g., vacations, sickness), and an assistant chief operator has not been designated, or, if designated, for any reason is unable to assume the duties of chief operator, the licensee shall designate another first-class radiotelephone operator as acting chief operator on a temporary basis. Within 3 days of the date such action is taken, the engineer in charge of the radio district in which the station is located shall be notified by the licensee by letter of the name and license number of the acting chief operator, and shall be notified by letter, again within 3 days of the date when the regularly designated chief operator returns to duty.

(5) The designated chief operator may serve as a routine duty transmitter operator at any station only to the extent that it does not interfere with the efficient discharge of his responsibilities as listed below.

(i) The inspection and maintenance of the transmitting system, including the antenna system and required monitoring equipment.

(ii) The accuracy and completeness of entries in the maintenance log.

(iii) The supervision and instruction of all other station operators in the performance of their technical duties.

(iv) A review of completed operating logs to determine whether technical operation of the station has been in accordance with the rules and terms of the station authorization. After review, the chief operator shall sign the log and indicate the date and time of such review. If the review of the operating logs indicates technical operation of the station is in violation of the rules or terms of the station authorization, he shall promptly initiate corrective action. The review of each day's operating logs shall be made within 24 hours, except that, if the chief operator is not on duty during a 24-hour period, the logs must be reviewed within 2 hours after his next appearance for duty. In any case, the time before review cannot exceed 72 hours.

(e) Subject to the conditions set forth in paragraphs (c) and (d) of this section, routine operation of the transmitting system may be performed by an operator holding a first-class radiotelegraph or radiotelephone license, a second-class radiotelegraph or radiotelephone license, or a third-class radiotelegraph or radiotelephone permit endorsed for broadcast station operation. Unless, however, performed under the immediate and personal supervision of an operator holding a first-class radiotelephone license, an operator holding a first-class radiotelegraph license, a second-class radiotelegraph or radiotelephone license, or a third-class radiotelegraph or radiotelephone permit endorsed for broadcast station operation, may make adjustment only of external controls, as follows:

(1) Those necessary to turn the transmitter on and off;

(2) Those necessary to compensate for voltage fluctuations in the primary power supply;

(3) Those necessary to maintain modulation levels of the transmitter within the prescribed limits.

(f) It is the responsibility of the station licensee to insure that each operator is fully instructed in the performance of all of the above adjustments as well as in other required duties, such as reading meters and making log entries. Printed step-by-step instructions for those adjustments which the lesser grade operator is permitted to make, and a tabulation or chart of upper and lower limiting values of parameters required to be observed and logged, shall be posted at the operating position. The emissions of the station shall be terminated immediately whenever the transmitting system is observed operating beyond the posted parameters, or in any other manner inconsistent with the rules or the station authorization and the above adjustments are ineffective in correcting the condition of improper operation and a first-class radiotelephone operator is not present.

(g) The operator on duty at the transmitter site or remote control point, may, at the discretion of the licensee and the chief operator, if any, be employed for other duties or for the operation of another radio station or stations in accordance with the class of operator's license which he holds and the rules and regulations governing such other stations: *Provided, however,* That such other duties shall not interfere with the proper operation of the transmitting system and keeping of required logs.

(h) At all FM broadcast stations, a complete inspection of the transmitting system and required monitoring equipment in use shall be made by an operator holding a first-class radiotelephone license at least once each calendar week. The interval between successive required inspections shall not be less than 5 days. This inspection shall include such tests, adjustments, and repairs as may be necessary to insure operation in conformance with the provisions of this subpart and the current station authorization.

§ 73.267 Operating power; determination and maintenance of.

(a) *Determination.* The operating power of each station shall be determined by either the direct or indirect method.

(1) Using the direct method, the power shall be measured at the output terminals of the transmitter while operating into a dummy load of substantially

zero reactance and a resistance equal to the transmission line characteristic impedance. The transmitter shall be unmodulated during this measurement. If electrical devices are used to determine the power output, such devices shall permit determination of this power to within an accuracy of ±5 percent of the power indicated by the full scale reading of the electrical indicating instrument of the device. If temperature and coolant flow indicating devices are used to determine the power output, such devices shall permit determination of this power to within an accuracy of 4 percent of measured average power output. During this measurement the direct plate voltage and current of the last radio stage and the transmission line meter shall be read and compared with similar readings taken with the dummy load replaced by the antenna. These readings shall be in substantial agreement.

(2) Using the indirect method, the operating power is the product of the plate voltage (E_p) and the plate current (I_p) of the last radio stage, and an efficiency factor, F, as follows:

$$\text{Operating power} = E_p \times I_p \times F$$

(3) The efficiency factor, F, shall be established by the transmitter manufacturer for each type of transmitter for which he submits data to the Commission, over the entire operating range of powers for which the transmitter is designed, and shall be shown in the instruction books supplied to the customer with each transmitter. In the case of composite equipment, the factor F shall be furnished to the Commission with a statement of the basis used in determining such factor.

(b) *Maintenance.* (1) The operating power shall be maintained as near as practicable to the authorized power and shall not be less than 90 percent nor greater than 105 percent of authorized power except as indicated in paragraph (c) of this section.

(2) When determined by the direct method, the operating power of the transmitter shall be monitored by a transmission line meter which reads proportional to the voltage, current, or power at the output terminals of the transmitter, the meter to be calibrated at intervals not exceeding 6 months. The calibration shall cover, as a minimum, the range from 90 to 105 percent of authorized power, and the meter shall provide clear indications which will permit maintaining the operating power within the prescribed tolerance or the meter shall be calibrated to read directly in power units.

(c) *Reduced power.* In the event it becomes technically impossible to operate with authorized power, the station may be operated with reduced power for a period of not more than 30 days without further authority from the Commission, *Provided,* That notification is sent to the Commission in Washington, D.C. not later than the 10th day of the lower power operation. In the event normal power is restored prior to the expiration of the 30 day period, the permittee or licensee will so notify the Commission in Washington, D.C. of this date. If causes beyond the control of the permittee or licensee prevent restoration of authorized power within the allowed period, informal written request shall be made to the Commission in Washington, D.C. no later than the 30th day for such additional time as may be deemed necessary.

§ 73.268 **Modulation.**

The percentage of modulation shall be maintained as high as possible consistent with good quality of transmission and good broadcast practice. In no case is it to exceed 100 percent on peaks of frequent recurrence. Generally, it should not be less than 85 percent on peaks of frequent recurrence; but where necessary to avoid objectionable loudness modulation may be reduced to whatever level is necessary, even if the resulting modulation is substantially less than 85 percent on peaks of frequent recurrence.

§ 73.269 **Frequency tolerance.**

The center frequency of each FM broadcast station shall be maintained within 2000 hertz of the assigned center frequency.

* * * * * * *

§ 73.274 **Remote control authorization.**

(a) An application to operate a station by remote control, to add a remote control point, or to change the location of a remote control point shall be made on FCC Form 301-A, except that:

(1) A request to operate a new station by remote control may be included in the application (FCC Form 301) for construction permit or modification of construction permit.

(2) A request to change a remote control point to a new studio location beyond the corporate limits of the community to which the station is assigned and at a point other than the authorized transmitter site may be included in the application (FCC Form 301) for authority to change the main studio location.

(3) No application need be filed to change a remote control point to an authorized main studio location within the corporate limits of the community to which the station is assigned or to its authorized transmitter site, or to delete a remote control point. However, any such change shall be reported to the Commission and to the Engineer in Charge of the radio district in which the station is located.

(b) An authorization for remote control will be issued only after a satisfactory showing has been made, including, among other things, the location of the remote control point(s).

§ 73.275 **Remote control operation.**

(a) Operation by remote control shall be subject to the following conditions:

(1) The equipment at the operating and transmitting positions shall be so installed and protected that it is not accessible to or capable of operation by persons other than those duly authorized by the licensee.

(2) The control circuits from the operating position to the transmitter shall provide positive on and off control and shall be such that open circuits, short circuits, grounds or other line faults will not actuate the transmitter and any fault causing loss of such control will automatically place the transmitter in an inoperative position.

(3) A malfunction of any part of the remote control system resulting in improper control shall be cause for the immediate cessation of operation by remote control. A malfunction of any part of the remote control system, resulting in inaccurate meter readings, shall be

cause for terminating operation by remote control no longer than 1 hour after the malfunction is detected.

(4) Control and monitoring equipment shall be installed so as to allow the licensed operator at the remote control point to perform all the functions in a manner required by the provisions of this part.

(5) Calibration of required indicating instruments at each remote control point shall be made against the corresponding instruments at the transmitter site at least once each calendar week. Results of calibration shall be entered in the maintenance log. Remote control point meters shall be calibrated to provide an indication within 2 percent of the corresponding instrument reading at the transmitter site. In no event shall a remote control meter be calibrated against another remote meter.

(6) All remote control meters shall conform with specifications required for the regular transmitter, antenna, and monitor meters.

(7) Meters with arbitrary scale divisions may be used provided that calibration charts or curves are provided at the transmitter remote control point showing the relationship between the arbitrary scales and the reading of the main meters.

(b) All stations, whether operating by remote control or direct control, shall be equipped so as to be able to follow the Emergency Action Notification procedures described in § 73.911.

§ 73.276 Extension meters.

The extension of required meters is permitted, without prior authorization of the Commission, upon compliance with each of the following:

(a) That the transmitter is in the same building as the normal operating location of the station's licensed operator and is no more than one floor above or below the normal operating location.

(b) That the path from the normal operating location to the transmitter is no longer than 100 feet and provides the operator with ready access to the transmitter.

(c) That the required extension meters and monitoring devices are sufficiently close to the operator's normal operating location that deviations from normal indications of such instruments can be observed from that location.

(d) That extension meters shall be installed for monitoring the direct plate voltage and current of the last radio stage and the transmission line radio frequency current, voltage, or power. The installation and operation thereof must comply with the same requirements prescribed by the rules for their corresponding regular meters.

(e) That each of the extension meters required in paragraph (d) of this section shall continuously sample the parameter for which it was installed and constantly indicate that parameter.

(f) That the extension meters required pursuant to paragraph (d) of this section are calibrated against their corresponding regular meters as often as necessary to insure their accuracy, but in no event less than once a week and;

(1) The results of such calibration shall be entered in the station's maintenance log.

(2) In no event shall an extension meter be calibrated against another extension or remote meter.

(3) Each extension meter shall be accurate within 2 percent of the value read on its corresponding regular meter.

(g) That the station's modulation monitor is installed at the same location as the extension meters: *Provided,* That, the modulation monitor may be installed at the transmitter if the extension meter location is equipped with a percentage modulation meter and peak indicating device which provide continuous and accurate indications of pertinent levels of total modulation.

(h) That in the event a malfunction of any component of the extension meter system causes inaccurate readings, the pertinent entries required in the station's operating log must be read and logged at the specified intervals from the meters located at the transmitter. If a malfunction affects extended indications of the modulation monitor, the licensee shall, pending repair or replacement, provide other suitable means for monitoring modulation at the extension meter location. When a malfunction is detected, an appropriate entry shall be made in the station's maintenance log, showing the date of occurrence and identifying the indicating device(s) affected. An entry, appropriately dated, shall also be made when repair or replacement is completed. If a malfunctioning component cannot be repaired or replaced within 60 days from the date faulty operation is detected, the Engineer in Charge of the radio district in which the station is located shall be notified and request made for such additional time as is needed to complete the necessary repair or replacement.

(i) That the transmitter is so installed and protected that it is not accessible to persons other than those duly authorized by the licensee.

§ 73.277 Permissible transmissions.

(a) No FM broadcast licensee or permittee shall enter into any agreement, arrangement or understanding, oral or written, whereby it undertakes to supply, or receives consideration for supplying, on its main channel a functional music, background music, or other subscription service (including storecasting) for reception in the place or places of business of any subscriber.

(b) The transmission (or interruption) of radio energy in the FM broadcast band is permissible only pursuant to a station license, program test authorization, Subsidiary Communications Authorization (SCA) or other specific authority therefor.

§ 73.281 General requirements relating to logs.

(a) The licensee or permittee of each FM broadcast station shall maintain program, operating, and maintenance logs as set forth in §§ 73.282, 73.283, and 73.284. Each log shall be kept by the station employee or employees (or contract operator) competent to do so, having actual knowledge of the facts required, who in the case of program and operating logs shall sign the appropriate log when starting duty, and again when going off duty.

(b) The logs shall be kept in an orderly and legible manner, in suitable form, and in such detail that the data required for the particular class of station concerned is readily available. Key letters or abbreviations may be used if proper meaning or explanation is contained elsewhere in the log. Each sheet shall be numbered and dated. Time entries shall be made in local time.

(c) No log or preprinted log or schedule which upon completion becomes a log, or portion thereof, shall be erased, obliterated, or willfully destroyed within the period of retention provided by the provisions of this part. Any necessary correction shall be made only pursuant to §§ 73.282, 73.283 and 73.284, and only by striking out the erroneous portion, or by making a corrective explanation on the log, or attachment to it as provided in those sections.

(d) Entries shall be made in the logs as required by §§ 73.282, 73.283, and 73.284. Additional information such as that needed for billing purposes or for the cuing of automatic equipment may be entered on the logs. Such additional information, so entered, shall not be subject to the restrictions and limitations in the Commission's rules on the making of corrections and changes in logs.

(e) The operating log and the maintenance log may be kept individually on the same sheet in one common log, at the option of the permittee or licensee.

§ 73.282 Program log.

(a) The following entries shall be made in the program log:

(1) *For each program.* (i) An entry identifying the program by name or title.

(ii) An entry of the time each program begins and ends. If programs are broadcast during which separately identifiable program units of a different type or source are presented, and if the licensee wishes to count such units separately, the beginning and ending time for the longer program need be entered only once for the entire program. The program units which the licensee wishes to count separately shall then be entered underneath the entry for a longer program, with the beginning and ending time of each such unit, and with the entry indented or otherwise distinguished so as to make it clear that the program unit referred to was broadcast within the longer program.

(iii) An entry classifying each program as to type, using the definitions set forth in NOTE 1 at the end of this section.

(iv) An entry classifying each program as to source, using the definitions set forth in NOTE 2 at the end of this section. (For network programs, also give name or initials of network, e.g., ABC, CBS, NBC, Mutual.)

(v) An entry for each program presenting a political candidate, showing the name and political affiliation of such candidate.

(2) *For commercial matter.* (i) An entry identifying (*a*) the sponsor(s) of the program; (*b*) the person(s) who paid for the announcement, or (*c*) the person(s) who furnished the materials or services referred to in § 73.289(d). If the title of a sponsored program includes the name of the sponsor, e.g., XYZ News, a separate entry for the sponsor is not required.

(ii) An entry or entries showing the total duration of commercial matter in each hourly time segment (beginning on the hour) or the duration of each commercial message (commercial continuity in sponsored programs, or commercial announcements) in each hour. See Note 5 at the end of this section for statement as to computation of commercial time.

(iii) An entry showing that the appropriate announcement(s) (sponsorship, furnishing material or services, etc.) have been made as required by Section 317 of the Communications Act and § 73.289. A checkmark (V) will suffice but shall be made in such a way as to indicate the matter to which it relates.

(3) *For public service announcements.* (i) An entry showing that a public service announcement (PSA) has been broadcast together with the name of the organization or interest on whose behalf it is made. See Note 4 following this section for definition of a public service announcement.

(4) *For other announcements.* (i) An entry of the time that each required station identification announcement is made (call letters and licensed location; see § 73.1201).

(ii) An entry for each announcement presenting a political candidate, showing the name and political affiliation of such candidate.

(iii) An entry for each announcement made pursuant to the local notice requirements of §§ 1.580 (pregrant) and 1.594 (designation for hearing) of this chapter, showing the time it was broadcast.

(iv) An entry showing that broadcast of taped, filmed, or recorded material has been made in accordance with the provisions of § 73.1208.

(b) Program log entries may be made either at the time of or prior to broadcast. A station broadcasting the programs of a national network which will supply it with all information as to such programs, commercial matter and other announcements for the composite week need not log such data but shall record in its log the time when it joined the network, the name of each network program broadcast, the time it leaves the network, and any nonnetwork matter broadcast required to be logged. The information supplied by the network, for the composite week which the station will use in its renewal application, shall be retained with the program logs and associated with the log pages to which it relates.

(c) No provision of this section shall be construed as prohibiting the recording or other automatic maintenance of data required for program logs. However, where such automatic logging is used, the licensee must comply with the following requirements:

(1) The licensee, whether employing manual or automatic logging or a combination thereof, must be able accurately to furnish the Commission with all information required to be logged;

(2) Each recording, tape, or other means employed shall be accompanied by a certificate of the operator or other responsible person on duty at the time or other duly authorized agent of the licensee, to the effect that it accurately reflects what was actually broadcast. Any information required to be logged which cannot be incorporated in the automatic process shall be maintained in a separate record which shall similarly be authenticated;

(3) The licensee shall extract any required information from the recording for the days specified by the Commission or its duly authorized representative and submit it in written log form, together with the underlying recording, tape, or other means employed.

(d) Program logs shall be changed or corrected only in the manner prescribed in § 73.281(c) and only in accordance with the following:

(1) *Manually kept log.* Where, in any program

log, or preprinted program log, or program schedule which upon completion is used as a program log, a correction is made before the person keeping the log has signed the log upon going off duty, such correction, no matter by whom made, shall be initialed by the person keeping the log prior to his signing of the log when going off duty, as attesting to the fact that the log as corrected is an accurate representation of what was broadcast. If corrections or additions are made on the log after it has been so signed, explanation must be made on the log or on an attachment to it, dated and signed by either the person who kept the log, the station program director or manager, or an officer of the licensee.

NOTE 1. *Program type definitions.* The definitions of the first eight types of programs (a) through (h) are intended not to overlap each other and will normally include all the various programs broadcast. Definitions (i) through (k) are subcategories and the programs classified thereunder will also be classified under one of the appropriate first eight types. There may also be further duplication within types (1) through (k); (e.g., a program presenting a candidate for public office, prepared by an educational institution, would be classified as Public Affairs (PA), Political (POL), and Educational Institution (ED).

(a) Agricultural programs (A) include market reports, farming, or other information specifically addressed, or primarily of interest, to the agricultural population.

(b) Entertainment programs (E) include all programs intended primarily as entertainment, such as music, drama, variety, comedy, quiz, etc.

(c) News programs (N) include reports dealing with current local, national, and international events, including weather and stock market reports; and when an integral part of a news program, commentary, analysis, and sports news.

(d) Public affairs programs (PA) include talks, commentaries, discussions, speeches, editorials, political programs, documentaries, forums, panels, round tables, and similar programs primarily concerning local, national, and international public affairs.

(e) Religious programs (R) include sermons or devotionals; religious news; and music, drama, and other types of programs designed primarily for religious purposes.

(f) Instructional programs (I) include programs (other than those classified under Agricultural, News, Public Affairs, Religious, or Sports) involving the discussion of, or primarily designed to further an appreciation or understanding of, literature, music, fine arts, history, geography, and the natural and social sciences; and programs devoted to occupational and vocational instruction, instruction with respect to hobbies, and similar programs intended primarily to instruct.

(g) Sports programs (S) include play-by-play and pre- or post-game related activities and separate programs of sports instruction, news or information (e.g., fishing opportunities, golfing instruction, etc.).

(h) Other programs (O) include all programs not falling within definitions (a) through (g).

(i) Editorials (EDIT) include programs presented for the purpose of stating opinions of the licensee.

(j) Political programs (POL) include those which present candidates for public office or which give expressions (other than in station editorials) to views on such candidates or on issues subject to public ballot.

(k) Educational Institution programs (ED) include any program prepared by, in behalf of, or in cooperation with, educational institutions, educational organizations, libraries, museums, PTA's, or similar organizations. Sports programs shall not be included.

NOTE 2. *Program source definitions.* (a) A local program (L) is any program originated or produced by the station, or for the production of which the station is primarily responsible, employing live talent more than 50 percent of the time. Such a program, taped or recorded for later broadcast, shall be classified as local. A local program fed to a network shall be classified by the originating station as local. All nonnetwork news programs may be classified as local. Programs primarily featuring records or transcriptions shall be classified as recorded (REC) even though a station announcer appears in connection with such material. However, identifiable units of such programs which are live and separately logged as such may be classified as local. (E.g., if during the course of a program featuring records or transcriptions a nonnetwork 2-minute news report is given and logged as a news program, the report may be classified as local.)

(b) A network program (NET) is any program furnished to the station by a network (national, regional or special). Delayed broadcasts of programs originated by networks are classified as networks.

(c) A recorded program (REC) is any program not otherwise defined in this Note including, without limitation, those using recordings, transcriptions, or tapes.

NOTE 3. *Definition of commercial matter* (CM) includes commercial continuity (network and nonnetwork) and commercial announcements (network and nonnetwork) as follows: (Distinction between continuity and announcements is made only for definition purposes. There is no need to distinguish between the two types of commercial matters when logging.)

(a) Commercial continuity (CC) is the advertising message of a program sponsor.

(b) A commercial announcement (CA) is any other advertising message for which a charge is made, or other consideration is received.

(1) Included are (i) "bonus spots"; (ii) trade-out spots, and (iii) promotional announcements of a future program where consideration is received for such an announcement or where such announcement identifies the sponsor of a future program beyond mention of the sponsor's name as an integral part of the title of the program. (E.g., where the agreement for the sale of time provides that the sponsor will receive promotional announcements, or when the promotional announcement contains a statement such as "Listen tomorrow for the—[name of program]—brought to you by—[sponsor's name]—.")

(2) Other announcements, including but not limited to the following, are not commercial announcements:

(i) Promotional announcements, except as heretofore defined in paragraph (b).

(ii) Station identification announcements for which no charge is made.

(iii) Mechanical reproduction announcements.

(iv) Public service announcements.

(v) Announcements made pursuant to § 73.289(d) that materials or services have been furnished as an inducement to broadcast a political program or a program involving the discussion of controversial public issues.

(vi) Announcements made pursuant to the local notice requirements of §§ 1.580 (pre-grant) and 1.594 (designation for hearing) of this chapter.

NOTE 4. *Definition of a public service announcement.* A public service announcement is an announcement for which no charge is made and which promotes programs, activities, or services of Federal, State, or local governments (e.g., recruiting, sales of bonds, etc.) or the programs, activities or services of nonprofit organizations (e.g., UGF, Red Cross Blood Donations, etc.), and other announcements regarded as serving community interests, excluding time signals, routine weather announcements and promotional announcements.

NOTE 5. *Computation of commercial time.* Duration of commercial matter shall be as close an approximation to the time consumed as possible. The amount of commercial time scheduled will usually be sufficient. It is not necessary, for example, to correct an entry of a 1-minute commercial to accommodate varying reading speeds even though the actual time consumed might be a few seconds more or less than the scheduled time. However, it is incumbent upon the licensee to ensure that the entry represents as close an approximation to the time actually consumed as possible.

§ 73.283 Operating log.

(a) The following entries shall be made in the operating log by the properly licensed operator in actual charge of the transmitting apparatus only:

(1) An entry of the time the station begins to supply power to the antenna and the time it stops.

(2) A notation of tests of the Emergency Broadcast System procedures pursuant to the requirements of Subpart G of this part and the appropriate station EBS checklist.

(3) An entry at the beginning of operation and at

intervals not exceeding 3 hours, of the following (actual readings observed prior to making any adjustments to the equipment and, when appropriate, an indication of corrections to restore parameters to normal operating values):

(i) Operating constants of last radio stage (total plate voltage and plate current).

(ii) RF transmission line meter reading, except when power is being determined by the indirect method.

(4) Any other entries required by the instrument of authorization or the provisions of this part.

(5) The entries required by § 17.49 (a), (b), and (c) of this chapter concerning daily observations of tower lights.

(b) Automatic devices accurately calibrated and with appropriate time, date and circuit functions may be utilized to record the entries in the operating log: *Provided*, That:

(1) They do not effect the operation of circuits or accuracy of indicating instruments of the equipment being recorded;

(2) The recording devices have an accuracy equivalent to the accuracy of the indicating instruments;

(3) The calibration is checked against the original indicators at least once a week and the results noted in the maintenance log;

(4) Provision is made to actuate automatically an aural alarm circuit located near the operator on duty if any of the automatic log readings are not within the tolerances or other requirements specified in the rules or instrument of authorization;

(5) Unless the alarm circuit operates continuously, devices which record each parameter in sequence must read each parameter at least once during each 10-minute period and clearly indicate the parameter being recorded;

(6) The automatic logging equipment is located at the remote control point if the transmitter is remotely controlled or at the transmitter location if the transmitter is directly controlled;

(7) The automatic logging equipment is located in the near vicinity of the operator on duty and inspected by him periodically during the broadcast day; and

(8) The indicating equipment conforms to the requirements of § 73.320 except that the scales need not exceed 2 inches in length. Arbitrary scales may not be used.

(c) In preparing the operating log, original data may be recorded in rough form and later transcribed into the log.

(d) Operating logs shall be changed or corrected only in the manner prescribed in § 73.281(c) and only in accordance with the following:

(1) *Manually kept log.* Any necessary corrections in a manually kept operating log shall be made only by the person making the original entry who shall make and initial each correction prior to signing the log when going off duty in accordance with § 73.281(a). If corrections or additions are made on the log after it has been so signed, explanation must be made on the log or on an attachment to it, dated and signed by either the operator who kept the log, the station technical supervisor or an officer of the licensee.

(2) *Automatic logging.* No automatically kept operating log shall be altered in any way after entries have been recorded. Any errors or omissions found in an automatically kept operating log shall be noted and explained in a memorandum signed by the operator on duty (who, under the provisions of paragraph (b)(7) of this section, is required to inspect the automatic equipment) or by the station technical supervisor or an officer of the licensee. Such memorandum shall be affixed to the original log in question.

(e) If required by § 73.265(d)(5)(iv), each completed operating log shall bear a signed notation by the station's chief operator of the results of the review of that log, and show the date and time of such review.

* * * * * * *

§ 73.285 Retention of logs.

Logs of FM broadcast stations shall be retained by the licensee or permittee for a period of 2 years: *Provided, however*, That logs involving communications incident to a disaster or which include communications incident to or involved in an investigation by the Commission and concerning which the licensee or permittee has been notified, shall be retained by the licensee or permittee until he is specifically authorized in writing by the Commission to destroy them: *Provided, further*, That logs incident to or involved in any claim or complaint of which the licensee or permittee has notice shall be retained by the licensee or permittee until such claim or complaint has been fully satisfied or until the same has been barred by statute limiting the time for the filing of suits upon such claims.

§ 73.286 Availability of logs and records.

The following shall be made available upon request by an authorized representative of the Commission:

(a) Program, operating and maintenance logs.

(b) Equipment performance measurements required by § 73.254.

* * * * * * *

§ 73.293 Subsidiary Communications Authorizations.

(a) A FM broadcast licensee or permittee may apply for a Subsidiary Communications Authorization (SCA) to provide limited types of subsidiary services on a multiplex basis. Permissible uses must fall within one or both of the following categories:

(1) Transmission of programs which are of a broadcast nature, but which are of interest primarily to limited segments of the public wishing to subscribe thereto. Illustrative services include: background music; storecasting; detailed weather forecasting; special time signals; and other material of a broadcast nature expressly designed and intended for business, professional, educational, religious, trade, labor, agricultural or other groups engaged in any lawful activity.

(2) Transmission of signals which are directly related to the operation of FM broadcast stations; for example: relaying of broadcast material to other FM and standard broadcast stations; remote cueing and order circuits; remote control telemetering functions associated with authorized STL operation, and similar uses.

(b) An application for an SCA shall be submitted on FCC Form 318. An applicant for SCA shall specify the particular nature and purpose of the proposed use.

If visual transmission of program material is contemplated (see § 73.310(c)), the application shall include certain technical information concerning the visual system, on which the Commission shall rely in issuing an SCA. If any significant change is subsequently made in the system, revised information shall be submitted. The technical information to be submitted is as follows:

(1) A full description of the visual transmission system.

(2) A block diagram of the system, as installed at the station, with all components, including filters, identified as to make and type. Response curves of all composite filters shall be furnished.

(3) The results of measurements which demonstrate that the subcarrier, when modulated by the visual signal, meets the requirements of § 73.319(e), and of such observations or measurements as may be necessary to show that signal components of appreciable strength are not produced outside of the band normally occupied by the FM station's emissions (see § 73.317(a)(12) and (13)). A description of the apparatus and techniques employed in these measurements and observations shall be furnished.

NOTE: Operation of an FM broadcast station to obtain the technical information necessary to support an application for an SCA for visual transmission shall be considered "* * * for experimental purposes in testing and maintaining apparatus * * *" and may be conducted without specific authorization from the Commission pursuant to § 73.262(a) of the rules. Tests may be conducted for this purpose during the period from 6 a.m. to midnight, with prior notification to the Commission and the Engineer in Charge of the radio district in which the station is located, subject to the provisions of § 73.262(b), (1), (2), and (3).

(c) SCA operations may be conducted without restriction as to time so long as the main channel is programmed simultaneously.

*   *   *   *   *

§ 73.295 Operation under Subsidiary Communications Authorizations.

(a) Operations conducted under a Subsidiary Communications Authorization (SCA) shall conform to the uses and purposes authorized by the Commission in granting the SCA application. Prior permission to engage in any new or additional activity must be obtained from the Commission pursuant to application therefor.

(b) Superaudible and subaudible tones and pulses may, when authorized by the Commission, be employed by SCA holders to activate and deactivate subscribers' multiplex receivers. The use of these or any other control techniques to delete main channel material is specifically forbidden.

(c) In all arrangements entered into with outside parties affecting SCA operation, the licensee or permittee must retain control over all material transmitted over the station's facilities, with the right to reject any material which it deems inappropriate or undesirable. Subchannel leasing agreements shall be reduced to writing, kept at the station, and made available for inspection upon request.

(d) The logging announcement, and other requirements imposed by §§ 73.282, 73.283, 73.284, 73.287, 73.289 and 73.1208 are not applicable to material transmitted on authorized subcarrier frequencies.

(e) To the extent that SCA circuits are used for the transmission of program material, each licensee or permittee shall maintain a daily program log in which a general description of the material transmitted shall be entered once during each broadcast day: *Provided, however,* That in the event of a change in the general description of the material transmitted, an entry shall be made in the SCA program log indicating the time of each such change and a description thereof.

(f) Each licensee or permittee shall maintain a daily operating log of SCA operation in which the following entries shall be made (excluding subcarrier interruptions of five minutes or less):

(1) Time subcarrier generator is turned on.
(2) Time modulation is applied to subcarrier.
(3) Time modulation is removed from subcarrier.
(4) Time subcarrier generator is turned off.

(g) The frequency of each SCA subcarrier shall be measured as often as necessary to ensure that it is kept at all times within 500 Hz of the authorized frequency. However, in any event, the measurement shall be made at least once each calendar month with not more than 40 days expiring between successive measurements.

(h) Program and operating logs for SCA operation may be kept on special columns provided on the station's regular program and operating log sheets.

(i) Technical standards governing SCA operation (§ 73.319) shall be observed by all FM broadcast stations engaging in such operation.

*   *   *   *   *

§ 73.297 Stereophonic broadcasting.

(a) FM broadcast stations may, without further authority, transmit stereophonic programs in accordance with the technical standards set forth in § 73.322: *Provided, however,* That the Commission in Washington, D.C. shall be notified within 10 days of the installation of type-accepted stereophonic transmission equipment or any change therein, and of the commencement of stereophonic programing.

(b) Each licensee or permittee engaging in stereophonic broadcasting shall measure the pilot subcarrier frequency as often as necessary to ensure that it is kept at all times within 2 Hz of the authorized frequency. However, in any event, the measurement shall be made at least once each calendar month with not more than 40 days expiring between successive measurements.

§ 73.298 Operation during emergency.

(a) When necessary to the safety of life and property and in response to dangerous conditions of a general nature, FM broadcast stations may, at the discretion of the licensee and without further Commission authority, transmit emergency weather warnings and other emergency information. Examples of emergency situations which may warrant either an immediate or delayed response by the licensee are: Tornadoes, hurricanes, floods, tidal waves, earthquakes, icing conditions, heavy snows, widespread fires, discharge of toxic gases, widespread power failures, industrial explosions, and civil disorders. Transmission of information concerning school closings and changes in schoolbus schedules resulting from any of these conditions, is appropriate. In addition, and if requested by responsible public officials, emergency point-to-point messages may be transmitted for the purpose of requesting or dispatching aid and assisting in rescue operations.

(b) When emergency operation is conducted under a State-Level EBS Operational Plan, the attention signal described in § 73.906 may be employed.

(c) Emergency operation shall be confined to the hours, frequencies, powers, and modes of operation specified in the license documents of the stations concerned.

(d) Any emergency operation undertaken in accordance with this section may be terminated by the Commission, if required in the public interest.

(e) Immediately upon cessation of an emergency during which broadcast facilities were used for the transmission of point-to-point messages under paragraph (a) of this section, a report in letter form shall be forwarded to the Commission in Washington, D.C., setting forth the nature of the emergency, the dates and hours of emergency operation, and a brief description of the material carried during the emergency period.

(f) If the Emergency Broadcast System (EBS) is activated at the National-Level while non-EBS emergency operation under this section is in progress, the EBS shall take precedence.

* * * * * * *

§ 73.310 Definitions.

(a) *Frequency modulation.*

Antenna height above average terrain. The average of the antenna heights above the terrain from 2 to 10 miles from the antenna for the eight directions spaced evenly for each 45 degrees of azimuth starting with True North. (In general, a different antenna height will be determined in each direction from the antenna. The average of these various heights is considered the antenna height above the average terrain. In some cases less than eight directions may be used. See § 73.313(d).) Where circular or elliptical polarization is employed, the antenna height above average terrain shall be based upon the height of the radiation center of the antenna which transmits the horizontal component of radiation.

Antenna power gain. The square of the ratio of the root-mean-square free space field strength produced at 1 mile in the horizontal plane, in millivolts per meter for 1 kilowatt antenna input power to 137.6 mv/m. This ratio should be expressed in decibels (dB). (If specified for a particular direction, antenna power gain is based on the field strength in that direction only.)

Center frequency. The term "center frequency" means:

(1) The average frequency of the emitted wave when modulated by a sinusoidal signal.

(2) The frequency of the emitted wave without modulation.

Effective radiated power. The term "effective radiated power" means the product of the antenna power (transmitter output power less transmission line loss) times (1) the antenna power gain, or (2) the antenna field gain squared. Where circular or elliptical polarization is employed, the term effective radiated power is applied separately to the horizontal and vertical components of radiation. For allocation purposes, the effective radiated power authorized is the horizontally polarized component of radiation only.

FM broadcast band. The band of frequencies extending from 88 to 108 megahertz, which includes those assigned to noncommercial educational broadcasting.

FM broadcast channel. A band of frequencies 200 kHz wide and designated by its center frequency. Channels for FM broadcast stations begin at 88.1 MHz and continue in successive steps of 200 kHz to and including 107.9 MHz.

FM broadcast station. A station employing frequency modulation in the FM broadcast band and licensed primarily for the transmission of radiotelephone emissions intended to be received by the general public.

Field strength. The electric field strength in the horizontal plane.

Free space field strength. The field strength that would exist at a point in the absence of waves reflected from the earth or other reflecting objects.

Frequency Modulation. A system of modulation where the instantaneous radio frequency varies in proportion to the instantaneous amplitude of the modulating signal (amplitude of modulating signal to be measured after pre-emphasis, if used) and the instantaneous radio frequency is independent of the frequency of the modulating signal.

Frequency swing. The instantaneous departure of the frequency of the emitted wave from the center frequency resulting from modulation.

Multiplex transmission. The term "multiplex transmission" means the simultaneous transmission of two or more signals within a single channel. Multiplex transmission as applied to FM broadcast stations means the transmission of facsimile or other signals in addition to the regular broadcast signals.

Percentage modulation. The ratio of the actual frequency swing to the frequency swing defined as 100 percent modulation, expressed in percentage. For FM broadcast stations, a frequency swing of ±75 kilohertz is defined as 100 percent modulation.

(b) *Stereophonic broadcasting.*

Cross-talk. An undesired signal occurring in one channel caused by an electrical signal in another channel.

FM stereophonic broadcast. The transmission of a stereophonic program by a single FM broadcast station utilizing the main channel and a stereophonic subchannel.

Left (or right) signal. The electrical output of a microphone or combination of microphones placed so as to convey the intensity, time, and location of sounds originating predominately to the listener's left (or right) of the center of the performing area.

Left (or right) stereophonic channel. The left (or right) signal as electrically reproduced in reception of FM stereophonic broadcasts.

Main channel. The band of frequencies from 50 to 15,000 hertz which frequency-modulate the main carrier.

Pilot subcarrier. A subcarrier serving as a control signal for use in the reception of FM stereophonic broadcasts.

Stereophonic separation. The ratio of the electrical signal caused in the right (or left) stereophonic channel to the electrical signal caused in the left (or right) stereophonic channel by the transmission of only a right (or left) signal.

Stereophonic subcarrier. A subcarrier having a

frequency which is the second harmonic of the pilot subcarrier frequency and which is employed in FM stereophonic broadcasting.

Stereophonic subchannel. The band of frequencies from 23 to 53 kilohertz containing the stereophonic subcarrier and its associated sidebands.

(c) *Visual transmission.* Transmissions of a broadcast nature on a subcarrier modulated with a signal of such characteristics as to permit its employment, in receivers of appropriate design, for visual presentation of the information so transmitted, e.g., on a viewing screen or a graphic record.

* * * * * * *

SUBPART C—NONCOMMERCIAL EDUCATIONAL FM BROADCAST STATIONS

* * * * * * *

§ 73.553 Modulation monitors.

(a) The licensee of each station licensed for transmitter power output above 10 watts shall have in operation, either at the transmitter or the extension meter location, or at the place the transmitter is controlled, a modulation monitor of a type approved by the Commission for non-multiplex operation: *Provided,* That: (1) If the station is engaged in stereophonic operation, as contemplated by § 73.596, the licensee shall have in operation a modulation monitor of a type approved by the Commission for monitoring stereophonic operation, and (2) if the station is engaged in operation with a Subsidiary Communications Authorization, as contemplated by § 73.595, the licensee shall have in operation a modulation monitor of a type approved by the Commission for monitoring SCA operation.

NOTE: Approved modulation monitors (non-multiplex, stereophonic, and SCA) are included on the Commission's "Radio Equipment List". Copies of this list are available for inspection at the Commission's offices in Washington, D.C., and at its field offices.

(b) In the event that the modulation monitor becomes defective, the station may be operated without the monitor pending its repair or replacement for a period not in excess of 60 days without further authority of the Commission: *Provided,* That:

(1) Appropriate entries shall be made in the maintenance log of the station showing the date and time the monitor was removed and restored to service.

(2) During the period when the station is operated without the modulation monitor, the licensee shall provide other suitable means for insuring that the modulation is maintained within the tolerance prescribed in § 73.568.

(c) If conditions beyond the control of the licensee prevent the restoration of the monitor to service within the above allowed period, informal request may be filed in accordance with § 1.549 of this chapter with the Engineer in Charge of the radio district in which the station is located for such additional time as may be required to complete repairs of the defective instrument.

(d) The licensee of each non-commercial educational FM broadcast station licensed for transmitter power output of 10 watts or less shall provide a percentage modulation indicator or a calibrated program level meter from which a satisfactory indication of the percentage of modulation of the transmitter can be determined.

* * * * * * *

§ 73.558 Indicating instruments.

(a) Each noncommercial FM broadcast station licensed for transmitter power above 10 watts shall be equipped with indicating instruments, which conform with the specifications set forth in § 73.320 for measuring the direct plate voltage and current of the last radio stage and the transmission line radio frequency current, voltage, or power.

(b) In the event that any one of these indicating instruments becomes defective when no substitute which conforms with the required specifications is available, the station may be operated without the defective instrument pending its repair or replacement for a period not in excess of 60 days: *Provided,* That:

(1) Appropriate entries shall be made in the maintenance log of the station showing the date and time the meter was removed from and restored to service.

(2) [Reserved]

(3) If the defective instrument is a plate voltmeter or plate ammeter in the last radio stage, the operating power shall be maintained by means of the radio frequency transmission line meter.

(c) If conditions beyond the control of the licensee prevent the restoration of the meter to service within the above allowed period, informal request may be filed in accordance with § 1.549 of this chapter with the Engineer in Charge of the radio district in which the station is located for such additional time as may be required to complete repairs of the defective instrument.

* * * * * * *

§ 73.564 Station and operator licenses; posting of.

(a) The station license and any other instrument of station authorization shall be posted in a conspicuous place and in such manner that all terms are visible at the place the licensee considers to be the principal control point of the transmitter. At all other control points listed on the station authorization, a photocopy of the station license and other instruments of station authorization shall be posted.

(b) The operator license, or Form 759 (Verification of Operator License or Permit), of each station operator employed full-time, part-time or via contract, shall be permanently posted and shall remain posted so long as the operator is employed by the licensee.

(1) The operator licenses shall be posted;

(i) Either at the transmitter or extension meter location; or

(ii) At the principal remote control point, if the station license authorizes operation by remote control.

(2) Posting of operator licenses shall be accomplished by affixing the license to the wall at the posting location, or enclosing in a binder, or inserting in folder and retaining at the posting location so that the licenses will be readily available and easily accessible at that location.

§ 73.565 Operator requirements.

(a) One or more operators holding a radio operator license or permit of a grade specified in this section shall be in actual charge of the transmitting system, and shall be on duty at the transmitter location, or at an authorized remote control point, or the position at which extension meters, as authorized pursuant to § 73.574 of this Subpart, are located. The transmitter and required monitors and metering equipment, or the required extension meters and monitoring equipment or the controls and required monitoring and metering equipment in an authorized remote control operation, shall be readily accessible to the licensed operator and located sufficiently close to the normal operating location that deviations from normal indications of required instruments can be observed from that location.

(b) With the exceptions set forth in paragraph (e) of this section, adjustments of the transmitting system, and inspection, maintenance, and required equipment performance measurements shall be performed only by an operator holding the class of license specified below, or during periods of operation when the transmitter is in the charge of an operator of the specified class, by or under the direction of a broadcast consultant regularly engaged in the practice of broadcast station engineering.

(1) A first-class radiotelephone operator license if the station is authorized to operate with transmitter power output of more than 1 kilowatt.

(2) A first-class or second-class radio-telephone operator license if the station is authorized to operate with transmitter power output of more than 10 watts but not in excess of 1 kilowatt.

(3) A first-class or second-class radiotelephone or radiotelegraph operator license if the station is authorized to operate with transmitter power output of not more than 10 watts.

(c) A noncommercial educational FM station with authorized transmitter output power not in excess of 25 kilowatts may employ first-class radiotelegraph or radiotelephone operators, second-class radiotelegraph or radiotelephone operators or operators with third-class radiotelegraph or radiotelephone permits endorsed for broadcast station operation, for the routine operation of the transmitting system, if the station has at least one operator of a class specified for this station's power category in paragraph (b) of this section, readily available at all times. This operator may be in full-time employment or, as an alternative, the licensee may contract in writing for the services, on a part-time basis, of one or more such operators. Signed contracts with part-time operators shall be kept in the files of the station and shall be made available for inspection upon request by an authorized representative of the Commission.

(d) A noncommercial educational FM station with authorized transmitter output power in excess of 25 kilowatts may employ first-class radiotelegraph operators, second-class radiotelegraph or radiotelephone operators, or operators with third-class radiotelegraph or radiotelephone permits endorsed for broadcast station operation for routine operation of the transmitting system if the station has in full-time employment at least one first-class radiotelephone operator and complies with the following:

(1) The station licensee shall designate one first-class radiotelephone operator in full-time employment as the chief operator who, together with the licensee, shall be responsible for the technical operation of the station. The licensee may also designate another first-class radiotelephone operator as assistant chief operator, who shall assume all responsibilities of the chief operator during periods of his absence. The station licensee shall notify the engineer in charge of the radio district in which the station is located of the name(s) and license number(s) of the operator(s) so designated. Such notification shall be made within 3 days of the date of such designation. A copy of the notification shall be posted with the license(s) of the designated operator(s).

(2) An operator designated as chief operator for one station may not be so designated concurrently at any other noncommercial educational FM broadcast station.

(3) The station licensee shall vest such authority in, and afford such facilities to the chief operator as may be necessary to insure that the chief operator's primary responsibility for the proper technical operation of the station may be discharged efficiently.

(4) At such times as the regularly designated chief operator is unavailable or unable to act as chief operator (e.g., vacations, sickness), and an assistant chief operator has not been designated, or, if designated, for any reason is unable to assume the duties of chief operator, the licensee shall designate another first-class radiotelephone operator as acting chief operator on a temporary basis. Within 3 days of the date such action is taken, the engineer in charge of the radio district in which the station is located shall be notified by the licensee by letter of the name and license number of the acting chief operator, and shall be notified by letter, again within 3 days of the date when the regularly designated chief operator returns to duty.

(5) The designated chief operator may serve as a routine duty transmitter operator at any station only to the extent that it does not interfere with the efficient discharge of his resopnsibilities as listed below.

(i) The inspection and maintenance of the transmitting system, including the antenna system and required monitoring equipment.

(ii) The accuracy and completeness of entries in the maintenance log.

(iii) The supervision and instruction of all other station operators in the performance of their technical duties.

(iv) A review of completed operating logs to determine whether technical operation of the station has been in accordance with the rules and terms of the station authorization. After review, the chief operator shall sign the log and indicate the date and time of such review. If the review of the operating logs indicates technical operation of the station is in violation of the rules or terms of the station authorization, he shall promptly initiate corrective action. The review of each day's operating logs shall be made within 24 hours, except that, if the chief operator is not on duty during a given 24 hour period, the logs must be reviewed within 2 hours after his next appearance for duty. In any case, the time before review cannot exceed 72 hours.

(e) Subject to the conditions set forth in paragraphs (c) and (d) of this section, routine operation of the transmitting system may be performed by an operator holding a first-class radiotelegraph license, a second-class radiotelegraph or radiotelephone license, or a third-class radiotelegraph or radiotelephone permit endorsed for broadcast operation. Unless, however, performed under the immediate and personal supervision of an operator of a class specified in paragrpah (b) of this section, the operator holding a lower class license may make adjustments only of external controls, as follows:

(1) Those necessary to turn the transmitter on and off;

(2) Those necessary to compensate for voltage fluctuations in the primary power supply;

(3) Those necessary to maintain modulation levels of the transmitter within prescribed limits.

(f) It is the responsibility of the station licensee to insure that each operator is fully instructed in the performance of all the above adjustments, as well as in other required duties, such as reading meters and making log entries. Printed step-by-step instruction for those adjustments which the lesser grade operator is permitted to make, and a tabulation or chart of upper and lower limiting values of parameters required to be observed and logged, shall be posted at the operating position. The emissions of the station shall be terminated immediately whenever the transmitting system is observed operating beyond the posted parameters, or in any other manner inconsistent with the rules or the station authorization, and the above adjustments are ineffective in correcting the condition of improper operation, and an operator of the class specified in paragraph (b) of this section is not present.

(g) The operator on duty at the transmitter site or remote control point, may, at the discretion of the licensee and the chief operator, if any, be employed for other duties or for the operation of another radio station or stations in accordance with the class of operator's license which he holds and the rules and regulations governing such other stations: *Provided, however,* That such other duties shall not interfere with the proper operation of the transmitting system and keeping of required logs.

(h) At all noncommercial educational FM broadcast stations, a complete inspection of the transmitting system and required monitoring equipment in use shall be made by an operator holding a first class radiotelephone license at least once each calendar week. The interval between successive required inspections shall not be less than 5 days. This inspection shall include such tests, adjustments, and repairs as may be necessary to insure operation in conformance with the provisions of this subpart and the current station authorization: *Provided,* That if the transmitter power output is in excess of 10 watts, but not greater than 1 kilowatt, an operator holding a second class radiotelephone license may perform the required inspection: *Provided, further,* That if the transmitter power output is 10 watts or less, no such weekly inspection need be made, although this shall in no way relieve such stations from the duty to operate in conformance with the provisions of this subpart and the current station authorization.

*　　*　　*　　*　　*　　*

§ 73.568 Modulation.

The percentage of modulation of all stations shall be maintained as high as possible consistent with good quality of transmission and good broadcast practice and in no case less than 85 percent or more than 100 percent on peaks of frequent recurrence during any selection which normally is transmitted at the highest level of the program under consideration.

*　　*　　*　　*　　*　　*

§ 73.581 General requirements relating to logs.

(a) The licensee or permittee of each noncommercial educational FM broadcast station shall maintain program, operating, and maintenance logs as set forth in §§ 73.582, 73.583, and 73.584. Each log shall be kept by the station employee or employees (or contract operator) competent to do so, having actual knowledge of the facts required, who in the case of program and operating logs shall sign the appropriate log when starting duty, and again when going off duty.

(b) The logs shall be kept in an orderly and legible manner, in suitable form, and in such detail that the data required for the particular class of station concerned is readily available. Key letters or abbreviations may be used if proper meaning or explanation is contained elsewhere in the log. Each sheet shall be numbered and dated. Time entries shall be made in local time.

(c) No log or preprinted log or schedule which upon completion becomes a log, or portion thereof shall be erased, obliterated, or willfully destroyed within the period of retention provided by the provisions of this part. Any necessary correction shall be made only pursuant to §§ 73.582, 73.583, and 73.584, and only by striking out the erroneous portion, or by making a corrective explanation on the log, or attachment to it as provided in those sections.

(d) Entries shall be made in the logs as required by §§ 73.582, 73.583, and 73.584. Additional information such as that needed for the cuing of automatic equipment may be entered on the logs. Such additional information, so entered shall not be subject to the restrictions and limitations in the Commission's rules on the making of corrections and changes in logs.

(e) The operating log and the maintenance log may be kept individually on the same sheet in one common log, at the option of the permittee or licensee.

§ 73.582 Program log.

(a) The following entries shall be made in the program log:

(1) An entry of the time each station identification announcement (call letters and location) is made.

(2) An entry briefly describing each program broadcast, such as "music", "drama", "speech", etc. together with the name or title thereof and the name of any donor announced pursuant to Section 73.289, with the time of the beginning and ending of the complete program. In addition, an entry reflecting the use of any taped, filmed, or recorded material, in accordance with the provisions of § 73.1208. If a speech is made by a political candidate, the name and political affiliations of such speaker shall be entered.

(b) No provision of this section shall be construed as prohibiting the recording or other automatic maintenance of data required for program logs. However, where such automatic logging is used, the licensee must comply with the following requirements:

(1) The licensee, whether employing manual or automatic logging or a combination thereof, must be able accurately to furnish the Commission with all information required to be logged;

(2) Each recording, tape, or other means employed shall be accompanied by a certificate of the operator or other responsible person on duty at the time or other duly authorized agent of the licensee, to the effect that

it accurately reflects what was actually broadcast. Any information required to be logged which cannot be incorporated in the automatic process shall be maintained in a separate record which shall be similarly authenticated.

(3) The licensee shall extract any required information from the recording for the days specified by the Commission or its duly authorized representative and submit it in written log form, together with the underlying recording, tape or other means employed.

(c) Program logs shall be changed or corrected only in the manner prescribed in § 73.581(c) and only in accordance with the following:

(1) *Manually kept log.* Where, in any program log, or preprinted program log, or program schedule which upon completion is used as a program log, a correction is made before the person keeping the log has signed the log upon going off duty, such correction, no matter by whom made, shall be initialed by the person keeping the log prior to his signing of the log when going off duty, as attesting to the fact that the log as corrected is an accurate representation of what was broadcast. If corrections or additions are made on the log after it has been so signed, explanation must be made on the log or on an attachment to it, dated and signed by either the person who kept the log, the station program director, or manager, or an officer of the licensee.

§ 73.583 Operating log.

(a) The following entries shall be made in the operating log by the properly licensed operator in actual charge of the transmitting apparatus only.

(1) An entry of the time the station begins to supply power to the antenna and the time it stops.

(2) A notation of tests of the Emergency Broadcast System procedures pursuant to the requirements of Subpart G of this part and the appropriate station EBS checklist.

(3) For each station licensed for transmitter output above 10 watts, an entry, at the beginning of operation and at intervals not exceeding 3 hours, of the following (actual readings observed prior to making any adjustments to the equipment, and when appropriate, an indication of corrections made to restore parameters to normal operating values):

(i) Operating constants of last radio stage (total plate voltage and plate current).

(ii) RF transmission line meter reading, except when power is being determined by the indirect method.

(4) Any other entries required by the instrument of authorization or the provisions of this part.

(5) The entries required by § 17.49 (a), (b), and (c) of this chapter concerning daily observations of tower lights.

(b) Automatic devices accurately calibrated and with appropriate time, date and circuit functions may be utilized to record the entries in the operating log: *Provided,* That:

(1) They do not affect the operation of circuits or accuracy of indicating instruments of the equipment being recorded;

(2) The recording devices have an accuracy equivalent to the accuracy of the indicating instruments;

(3) The calibration is checked against the original indicators at least once a week and the results noted in the maintenance log;

(4) Provision is made to actuate automatically an aural alarm circuit located near the operator on duty if any of the automatic log readings are not within the tolerances or other requirements specified in the rules or instrument of authorization;

(5) Unless the alarm circuit operates continuously, devices which record each parameter in sequence must read each parameter at least once during each 10-minute period and clearly indicate the parameter being recorded;

(6) The automatic logging equipment is located at the remote control point if the transmitter is remotely controlled, or at the transmitter location if the transmitter is manually controlled;

(7) The automatic logging equipment is located in the near vicinity of the operator on duty and is inspected by him periodically during the broadcast day; and

(8) The indicating equipment conforms to the requirements of § 73.320 except that the scales need not exceed 2 inches in length. Arbitrary scales may not be used.

(c) In preparing the operating log, original data may be recorded in rough form and later transcribed into the log.

(d) Operating logs shall be changed or corrected only in the manner prescribed in § 73.581(c) and only in accordance with the following:

(1) *Manually kept log.* Any necessary corrections in a manually kept operating log shall be made only by the person making the original entry who shall make and initial each correction prior to signing the log when going off duty in accordance with § 73.581(a). If corrections or additions are made on the operating log after it has been so signed explanation must be made on the log or on an attachment to it, dated and signed by either the operator who kept the log, the station technical supervisor or an officer of the licensee.

(2) *Automatic logging.* No automatically kept operating log shall be altered in any way after entries have been recorded. Any errors or omissions found in an automatically kept operating log shall be noted and explained in a memorandum signed by the operator on duty (who, under the provisions of paragraph (b)(7) of this section, is required to inspect the automatic equipment) or by the station technical supervisor or an officer of the licensee. Such memorandum shall be affixed to the original log in question.

(e) If required by § 73.565(d)(5)(iv), each completed operating log shall bear a signed notation by the station's chief operator of the results of the review of that log, and show the date and time of such review.

* * * * * * *

§ 73.586 Availability of logs and records.

The following shall be made available upon request by an authorized representative of the Commission:

(a) Program, operating and maintenance logs.

(b) Equipment performance measurements required by § 73.554.

* * * * * * *

§ 73.593 Subsidiary Communications Authorizations.

(a) A noncommercial educational FM broadcast

licensee or permittee may apply for a Subsidiary Communications Authorization (SCA) to provide limited types of subsidiary service on a multiplex basis. Any use of SCA by such licensee or permittee must be consistent with the limitation on the purpose and operation of noncommercial educational FM stations contained in § 73.503 : *Provided,* That uses permitted under this paragraph will not be considered "commercial" as long as no consideration for such use (other than the furnishing of the material transmitted and/or payment of line charges) is received by the licensee, directly or indirectly, and no commercial announcements or references are contained in the material transmitted under the SCA. Permissible uses must fall within one or both of the following categories:

(1) Transmission of programs which are nonmerical and in furtherance of an educational purpose, and which are of a broadcast nature but of interest primarily to limited segments of the station's audience. Typical services may include: programs for presentation in classrooms; programs designed for specific professional groups, such as doctors, lawyers, and engineers; programs intended to serve the special needs and interests of the aged, the handicapped, particular social or ethnic groups, and for those in a specific trade or sharing a common interest or hobby; programs for individualized remedial or advanced learning needs; and any use permitted for a commercial FM station under § 73.293(a)(1), subject to the prohibition against commercial operation and the limitation as to purpose contained in this section and in § 73.503, such limitation especially including those non-instructional services customarily provided by commercial firms. Users permitted under this subparagraph will not be considered "commercial," when charges are made for the service rendered, under the circumstances and subject to the conditions set forth hereunder:

(i) A per-course, per-session, per-seminar, per-pupil or other appropriate fee is charged for formal or informal instructional material, presented by, with or for a bona fide educational institution. Payment of the fee shall be made to the noncommercial educational FM station or to the educational institution; such fee may include, in addition to the station expenses detailed in subdivision (iii), of this subparagraph, the usual tuition charged for similar material presented by other means.

(ii) A charge is made for a program or series of programs, informational or generally instructional in nature, intended to meet the special needs and interests of one or more of the groups the station is authorized to serve under its SCA. Payment of the charge shall be made to the noncommercial educational FM station.

(iii) Payments retained by the station shall total no more than the approximate cost of conducting the SCA operation (including purchase or lease of equipment, course material, personnel services, etc.) and the general overhead and operational costs attributable to such operation.

(iv) A noncommercial educational FM station offering program material subject to fee or other charge shall clearly indicate in any broadcast or printed solicitation to prospective enrollees whether the material falls into category subdivision (i) or (ii), of this subparagraph, so that informational and general educational materials are not represented as formal instructional or institutional credit programs.

(2) Transmission of signals which are directly related to the operation of FM broadcast stations; for example, relaying of broadcast material to other broadcast stations, remote cueing and order circuits, remote control telemetering functions associated with authorized STL operation, and similar uses.

(b) An application for an SCA shall be submitted on FCC Form 318. An applicant for SCA shall specify the particular nature and purpose of the proposed use. If visual transmission of program material is contemplated (see § 73.310(c)), the application shall include certain technical information concerning the visual system, on which the Commission shall rely in issuing an SCA. If any significant change is subsequently made in the system, revised information shall be submitted. The technical information to be submitted is as follows:

(1) A full description of the visual transmission system.

(2) A block diagram of the system, as installed at the station, with all components, including filters, identified as to make and type. Response curves of all composite filters shall be furnished.

(3) The results of measurements which demonstrate that the subcarrier, when modulated by the visual signal, meets the requirements of § 73.319(e), and of such observations or measurements as may be necessary to show that signal components of appreciable strength are not produced outside of the band normally occupied by the FM station's emissions (see § 73.317(a)(12) and (13)). A description of the apparatus and techniques employed in these measurements and observations shall be furnished.

NOTE: Operation of an FM broadcast station to obtain the technical information necessary to support an application for an SCA for visual transmission shall be considered "* * * for experimental purposes in testing and maintaining apparatus * * *" and may be conducted without specific authorization from the Commission pursuant to § 73.262(a) of the rules. Tests may be conducted for this purpose during the period from 6 a.m. to midnight, with prior notification to the Commission and the Engineer in Charge of the radio district in which the station is located, subject to the provisions of § 73.562(b), (1), (2), and (3).

(c) SCA operations may be conducted without restriction as to time so long as the main channel is programmed simultaneously.

* * * * * * *

§ 73.595 **Operation under Subsidiary Communications Authorizations.**

(a) Operations conducted under a Subsidiary Communications Authorization (SCA) shall conform to the uses and purposes authorized by the Commission in granting the SCA application. Prior permission to engage in any new or additional activity must be obtained from the Commission pursuant to application therefor.

(b) Superaudible and subaudible tones and pulses may, when authorized by the Commission, be employed by SCA holders to activate and deactivate subscribers' multiplex receivers. The use of these or any other control techniques to delete main channel material is specifically forbidden.

(c) In all arrangements entered into with outside

parties affecting SCA operation, the licensee or permittee must retain control over all material transmitted over the station's facilities, with the right to reject any material which it deems inappropriate or undesirable. Subchannel leasing arrangements shall be reduced to writing, kept at the station, and made available for inspection upon request from the Commission.

(d) The logging, announcement, and other requirements imposed by §§ 73.582, 73.583, 73.584, 73.587, and 73.1208 are not applicable to material transmitted on authorized subcarrier frequencies.

(e) To the extent that SCA circuits are used for the transmission of program material, each licensee or permittee shall maintain a daily program log in which a general description of the material transmitted shall be entered once during each broadcast day: *Provided, however,* That in the event of a change in the general description of the material transmitted, an entry shall be made in the SCA program log indicating the time of each such change and a description thereof.

(f) Each licensee or permittee shall maintain a daily operating log of SCA operation in which the following entries shall be made (excluding subcarrier interruptions of five minutes or less):

(1) Time subcarrier generator is turned on.
(2) Time modulation is applied to subcarrier.
(3) Time modulation is removed from subcarrier.
(4) Time subcarrier generator is turned off.

(g) The frequency of each SCA subcarrier shall be measured as often as necessary to ensure that it is kept at all times within 500 Hz of the authorized frequency. In any event, however, SCA subcarrier frequencies shall be measured in accordance with the following schedule:

(1) For stations authorized to operate with transmitter power in excess of 10 watts, each SCA subcarrier frequency shall be measured at least once each calendar month with not more than 40 days expiring between successive measurements.

(2) For stations authorized to operate with transmitter power of 10 watts or less, each SCA subcarrier frequency will be measured:

(i) When the SCA subcarrier generator is initially installed;
(ii) At any time the frequency determining elements of the SCA subcarrier generator are changed;
(iii) At any time the licensee may have reason to believe the SCA subcarrier frequency is not within the frequency tolerance prescribed by the Commission's rules.

(h) Program and operating logs for SCA operation may be kept on special columns provided on the station's regular program and operating log sheets.

(i) Technical standards governing SCA operation (§ 73.319) shall be observed by all FM broadcasting stations engaging in such operation.

§ 73.596 Stereophonic broadcasting.

(a) Noncommercial educational FM broadcast stations may, without further authority, transmit stereophonic programs in accordance with the technical standards set forth in § 73.322: *Provided, however,* That the Commission, in Washington, D.C., shall be notified within 10 days, of the installation of type-accepted stereophonic transmission equipment or any change therein, and of the commencement of stereophonic programming.

(b) Each licensee or permittee engaging in stereophonic broadcasting shall measure the pilot subcarrier frequency as often as necessary to ensure that it is kept at all times within 2 Hz of the authorized frequency. In any event, however, the stereo-pilot subcarrier frequency shall be measured in accordance with the following schedule:

(i) For stations authorized to operate with transmitter power in excess of 10 watts, the pilot subcarrier frequency shall be measured at least once each calendar month with not more than 40 days expiring between successive measurements.

(2) For stations authorized to operate with transmitter power of 10 watts or less, the pilot subcarrier frequency shall be measured:

(i) When the stereo-pilot subcarrier generator is initially installed;
(ii) At any time the frequency determining elements of the stereo-pilot subcarrier generator are changed;
(iii) At any time the licensee may have reason to believe the stereo-pilot subcarrier frequency is not within the frequency tolerance prescribed by the Commission's rules.

* * * * * * *

SUBPART G—EMERGENCY BROADCAST SYSTEM

* * * * * * *

§ 73.903 Emergency Broadcast System (EBS).

The EBS is composed of AM, FM and TV broadcast stations and non-government industry entities operating on a voluntary, organized basis during emergencies at National, State or Operational (Local) Area levels.

* * * * * * *

§ 73.919 Non-participating Station.

This is a broadcast station which is not voluntarily participating in the National-Level EBS and does not hold an EBS Authorization. Such stations are required to remove their carriers from the air and monitor for emergency action termination in accordance with the instructions in the EBS Checklist for Non-Participating Stations.

* * * * * * *

§ 73.932 Radio Monitoring Requirement.

(a) To insure effective off-the-air monitoring (§ 73.-931(a)(3)) all broadcast station licensees must install and operate, during their hours of broadcast operation, equipment capable of receiving Emergency Action Notifications and Terminations transmitted by other radio broadcast stations. This equipment must be maintained in operative condition, including arrangements for human listening watch or automatic alarm devices, and shall have its termination at each transmitter control point. Where more than one broadcast transmitter is controlled from a common point by the same operator, only one receiver is required at that point.

(b) Off-the-air monitoring assignment of each broadcast station is specified in the State EBS Operational Plan.

(c) Prior to commencing routine operation or originating any emissions under program test, equipment test, experimental, or other authorizations or for any

other purpose, licensees or permittees shall first ascertain whether an EAN message has been released by any one or all of the following methods:
(1) Monitor the radio and TV network facilities.
(2) Check the Radio Press Wire Service (AP/UPI).
(3) Monitor the Primary Station and/or the Primary Relay Station for your Operational (Local) Area.
If so, operation shall be in accordance with this subpart of the rules.

* * * * * * *

§ 73.961 **Tests of the Emergency Broadcast System Procedures.**

* * * * * * *

(c) Weekly Off-The-Air Monitor Tests will be conducted by all AM, FM, and TV broadcast stations once each week between the hours of 8:30 a.m. and local sunset. These tests will be conducted in accordance with procedures set forth in the EBS Checklist furnished to all broadcast stations.

* * * * * * *

Subpart H—Rules Applicable in Common to Broadcast Stations

* * * * * * *

§ 73.1201 Station identification.

(a) *When regularly required.* Broadcast station identification announcements shall be made: (1) At the beginning and ending of each time of operation, and (2) hourly, as close to the hour as feasible, at a natural break in program offerings. Television broadcast stations may make these announcements visually or aurally.

(b) *Content.* (1) Official station identification shall consist of the station's call letters immediately followed by the name of the community or communities specified in its license as the station's location.

(2) When given specific written authorization to do so, a station may include in its official station identification the name of an additional community or communities, but the community to which the station is licensed must be named first.

(3) A licensee shall not in any identification announcements, promotional announcements or any other broadcast matter either lead or attempt to lead the station's audience to believe that the station has been authorized to identify officially with cities other than those permitted to be included in official station identifications under subparagraphs (1) and (2) of this paragraph.

NOTE: Commission interpretations of this paragraph may be found in a separate Public Notice issued Oct. 30, 1967, entitled "Examples of Application of Rule Regarding Broadcast of Statements Regarding a Station's Licensed Location." (FCC 67-1132; 10 FCC 2d 407).

(c) *Channel*—(1) *General.* Except as otherwise provided in this paragraph, in making the identification announcement the call letters shall be given only on the channel identified thereby.

(2) *Simultaneous AM-FM broadcasts.* If the same licensee operates an FM broadcast station and a standard broadcast station and simultaneously broadcasts the same programs over the facilities of both such stations, station identification announcements may be made jointly for both stations for periods of such simultaneous operation. If the call letters of the FM station do not clearly reveal that it is an **FM** station, the joint announcement shall so identify it.

* * * * * * *

§ 73.1205 **Fraudulent billing practices.**

No licensee of a standard, FM, or television broadcast station shall knowingly issue to any local, regional or national advertiser, advertising agency, station representative, manufacturer, distributor, jobber, or any other party, any bill, invoice, affidavit or other document which contains false information concerning the amount actually charged by the licensee for the broadcast advertising for which such bill, invoice, affidavit or other document is issued, or which misrepresents the nature or content of such advertising, or which misrepresents the quantity of advertising actually broadcast (number or length of advertising messages) or the time of day or date at which it was broadcast. Licensees shall exercise reasonable diligence to see that their agents and employees do not issue any documents which would violate this section if issued by the licensee.

* * * * * * *

§ 73.1206 Broadcast of telephone conversations.

Before recording a telephone conversation for broadcast, or broadcasting such a conversation simultaneously with its occurrence, a licensee shall inform any party to the call of the licensee's intention to broadcast the conversation, except where such party is aware, or may be presumed to be aware from the circumstances of the conversation, that it is being or likely will be broadcast. Such awareness is presumed to exist only when the other party to the call is associated with the station (such as an employee or part-time reporter), or where the other party originates the call and it is obvious that it is in connection with a program in which the station customarily broadcasts telephone conversations.

§ 73.1207 Rebroadcast

(a) The term "rebroadcast" means reception by radio of the programs of a radio station, and the simultaneous or subsequent retransmission of such programs by a broadcast station.

NOTE 1: As used in § 73.1207 "program" includes any complete program or part thereof.

NOTE 2: The transmission of a program from its point of origin to a broadcast station entirely by common carrier facilities, whether by wire line or radio, is not considered a rebroadcast.

(b) No broadcasting station shall rebroadcast the program, or any part thereof of another U.S. broadcasting station without the express authority of the originating station. A copy of the written consent of the licensee originating the program shall be kept by the licensee of the station rebroadcasting such program and shall be made available to the Commission upon request. Stations originating emergency communications under a Detailed State EBS Operational Plan, shall be deemed to have conferred rebroadcast authority on other participating stations. The broadcasting of a program relayed by a remote pickup broadcast station (§ 74.401 of this chapter) is not considered a rebroadcast.

* * * * * * *

§ 73.1208 Broadcast of taped, filmed, or recorded material.

(a) Any taped, filmed or recorded program material in which time is of special significance, or by which an affirmative attempt is made to create the impression that it is occurring simultaneously with the broadcast, shall be announced at the beginning as taped, filmed or recorded. The language of the announcement shall be clear and in terms commonly understood by the public. For television stations, the announcement may be made visually or aurally.

(b) Taped, filmed, or recorded announcements which are of a commercial, promotional or public service nature need not be identified as taped, filmed or recorded.

* * * * * *

§ 73.1211 Broadcast of lottery information.

(a) No licensee of an AM, FM or television broadcast station, except as in paragraph (c) of this section, shall broadcast any advertisement of or information concerning any lottery, gift enterprise, or similar scheme, offering prizes dependent in whole or in part upon lot or chance, or any list of the prizes drawn or awarded by means of any such lottery, gift enterprise or scheme, whether said list contains any part or all of such prizes. (18 U.S.C. 1304, 62 Stat. 763).

(b) The determination whether a particular program comes within the provisions of paragraph (a) of this section depends on the facts of each case. However, the Commission will in any event consider that a program comes within the provisions of paragraph (a) of this section if in connection with such program a prize consisting of money or other thing of value is awarded to any person whose selection is dependent in whole or in part upon lot or chance, if as a condition of winning or competing for such prize, such winner or winners are required to furnish any money or other thing of value or are required to have in their possession any product sold, manufactured, furnished or distributed by a sponsor of a program broadcast on the station in question. (See 21 FCC 2d 846).

(c) The provisions of paragraphs (a) and (b) of this section shall not apply to an advertisement, list of prizes or other information concerning a lottery conducted by a State acting under authority of State law when such information is broadcast: (1) by a radio or television broadcast station licensed to a location in that State, or (2) by a radio or television broadcast station licensed to a location in an adjacent State which also conducts such a lottery. (18 U.S.C. 1307, 88 Stat. 1916). ,

(d) For the purposes of paragraph (c) of this section, "lottery" means the pooling of proceeds derived from the sale of tickets or chances and allotting those proceeds or parts thereof by chance to one or more chance takers or ticket purchasers. It does not include the placing or accepting of bets or wagers on sporting events or contests.

NOTE.—Pursuant to the exemption set out in paragraph (c) of this section, a broadcast station licensed to a location in a State that conducts a State Lottery may broadcast advertisements of or information concerning such lottery in its State of license and advertisements of or information concerning such lotteries conducted in any adjacent State. (See 18 U.S.C. 1307, FCC 75). The exemption would, for example, permit a broadcast station licensed to a location in New York, which now conducts a lawful State Lottery, to broadcast advertisements of or information concerning the New York State Lottery as well as the lawful State Lotteries of Massachusetts, Connecticut, New Jersey and Pennsylvania, since these States are adjacent to New York, and also conduct a State Lottery. The exemption, however, would not permit a broadcast station licensed to a location in New York to broadcast information concerning the Maine or Michigan State lotteries since those States are not adjacent States to New York. Nor would the exemption permit a station licensed to a location in Virginia to broadcast information concerning the Maryland State Lottery, since although Virginia is adjacent to Maryland, Virginia does not conduct a State lottery.

§ 73.1212 Sponsorship identification; list retention; related requirements.

(a) When a broadcast station transmits any matter for which money, service, or other valuable consideration is either directly or indirectly paid or promised to, or charged or accepted by such station, the station, at the time of the broadcast, shall announce (1) that such matter is sponsored, paid for, or furnished, either in whole or in part, and (2) by whom or on whose behalf such consideration was supplied: *Provided, however,* That "service or other valuable consideration" shall not include any service or property furnished either without or at a nominal charge for use on, or in connection with, a broadcast unless it is so furnished in consideration for an identification of any person, product, service, trademark, or brand name beyond an identification reasonably related to the use of such service or property on the broadcast.

(i) For the purposes of this section, the term "sponsored" shall be deemed to have the same meaning as "paid for."

(b) The licensee of each broadcast station shall exercise reasonable dilligence to obtain from its employees, and from other persons with whom it deals directly in connection with any matter for broadcast, information to enable such licensee to make the announcement required by this section.

(c) In any case where a report has been made to a broadcast station as required by section 508 of the Communications Act of 1934, as amended, of circumstances which would have required an announcement under this section had the consideration been received by such broadcast station, an appropriate announcement shall be made by such station.

(d) In the case of any political broadcast matter or any broadcast matter involving the discussion of a controversial issue of public importance for which any film, record, transcription, talent, script, or other material or service of any kind is furnished, either directly or indirectly, to a station as an inducement for broadcasting such matter, an announcement shall be made both at the beginning and conclusion of such broadcast on which such material or service is used that such film, record, transcription, talent, script, or other material or service has been furnished to such station in connection with the transmission of such broadcast matter: *Provided, however,* That in the case of any broadcast of 5 minutes' duration or less, only one such announcement need be made either at the beginning or conclusion of the broadcast.

(e) The announcement required by this section shall, in addition to stating the fact that the broadcast matter was sponsored, paid for or furnished, fully and fairly disclose the true identity of the person or persons, or corporation, committee, association or other unincorporated group, or other entity by whom or on whose behalf such payment is made or promised, or from whom or on whose behalf such services or other valuable consideration is received, or by whom the material or services referred to in paragraph (d) of this section are furnished. Where an agent or other person or entity contracts or otherwise makes arrangements with a station on behalf of another, and such fact is known or by the exercise of reasonable diligence, as specified in paragraph (b) of this section, could be known to the station, the announcement shall disclose the identity of the person or persons or entity on whose behalf such agent is acting instead of the name of such agent. Where the the material broadcast is political matter or matter involving the discussion of a controversial issue of public importance and a corporation, committee, association or other unincorporated group, or other entity is paying for or furnishing the broadcast matter, the station shall, in addition to making the announcement required by this section, require that a list of the chief executive officers or members of the executive committee or of the board of directors of the corporation, committee, association or other unincorporated group, or other entity shall be made available for public inspection at the location specified by the licensee under § 1.526 of this Chapter. If the broadcast is originated by a network, the list may, instead, be retained at the headquarters office of the network or at the location where the originating station maintains its public inspection file under § 1.526 of this chapter. Such lists shall be kept and made available for a period of two years.

(f) In the case of broadcast matter advertising commercial products services, an announcement stating the sponsor's corporate or trade name, or the name of the sponsor's product, when it is clear that the mention of the name of the product constitutes a sponsorship identification, shall be deemed sufficient for the purpose of this section and only one such announcement need be made at any time during the course of the broadcast.

(g) The announcement otherwise required by section 317 of the Communications Act of 1934, as amended, is waived with respect to the broadcast of "want ad" or classified advertisements sponsored by an individual. The waiver granted in this paragraph shall not extend to a classified advertisement or want ad sponsorship by any form of business enterprise, corporate or otherwise. Whenever sponsorship announcements are omitted pursuant to this paragraph, the licensee shall observe the following conditions:

(1) Maintain a list showing the name, address, and (where available) the telephone number of each advertiser;

(2) Attach the list to the program log for the day when such broadcast was made; and

(3) Make this list available to members of the public who have a legitimate interest in obtaining the information contained in the list.

(h) Any announcement required by section 317(b) of the Communications Act of 1934, as amended, is waived with respect to feature motion picture film produced initially and primarily for theatre exhibition.

NOTE: The waiver heretofore granted by the Commission in its Report and Order adopted November 16, 1960 (FCC 60-1369; 40 F.C.C. 95), continues to apply to programs filmed or recorded on or before June 20, 1963, when § 73.654, the predecessor television rule, went into effect.

(i) Commission interpretations in connection with the provisions of the sponsorship identification rules are contained in the Commission's Public Notice, entitled "Applicability of Sponsorship Identification Rules," dated May 6, 1963 (40 F.C.C. 141), as modified by Public Notice, dated April 21, 1975 (FCC 75-418). Further interpretations are printed in full in various volumes of the Federal Communications Commission Reports.

§ 73.1213 Antenna structure, marking and lighting.

(a) The provisions of Part 17 of this Chapter (Construction, Marketing and Lighting of Antenna Structures), require certain antenna structures be painted and/or lighted in accordance with the provisions of that Part. (See §§ 17.47 through 17.56.)

(b) The licensee or permittee of an AM, FM, or TV broadcast station, if the sole occupant of the antenna and/or the antenna supporting structure, is responsible for conforming to the requirements of §§ 17.47 through 17.56 of this chapter.

(c) If a common tower is used for antenna and/or antenna supporting purposes by more than one licensee or permittee of an AM, FM, or TV station or by one or more such licensees or permittees of any other service, each licensee or permittee shall be responsible for painting and lighting the structure when obstruction marking and lighting are required by Commission rules. However, each such licensee or permittee utilizing a common tower may, with the approval of the Commission in Washington, designate one of the licensees or permittees as responsible for painting and lighting the structure. Pursuant to Commission approval, such designated licensee or permittee shall be solely responsible for conforming to all Commission requirements of Part 17 of this Chapter regarding obstruction marking and lighting of antenna structures. (See §§ 17.47 through 17.56.) Requests for such approval shall be submitted in letter form, accompanied by copies of agreements between all participating licensees or permittees. A copy of the agreement and the Commission approval must be retained in each licensee's or permittee's station file, available for inspection by FCC representatives. In the event of default by the designated licensee of his responsibility, each of the licensees or permittees shall again be individually responsible for conforming to the requirements of the rules, pending Commission approval of a new agreement.

APPENDIX V

Information Concerning Commercial Radio Operator Licenses And Permits

THE APPLICATION

(1) An applicant must normally be a citizen or national of the United States. Under certain circumstances U.S. nationality may be waived for:
  1) *Alien aircraft pilots, and* 2) *Citizens of a U.S. Trust Territory.*
(2) Except for Restricted Radiotelephone Operator Permit, submit in advance a completed application Form 756 and appropriate fee to the office which will administer the examination. If examination is to be taken at an FCC office the forms may be obtained and completed at the time of the examination. To request a Restricted Radiotelephone Operator Permit for which no examination is required, submit FCC application Form 753 and fee by mail to Commission's office at Gettysburg, Pa. 17325.
(3) The Commission has not established any age limit for applicants who wish to obtain commercial radio operator licenses, except that radiotelegraph first-class operator licenses may not be issued to applicants under twenty-one (21) years of age, applicants for examination for an Aircraft Radiotelegraph Endorsement must be at least eighteen (18) years of age, and applicants for the Restricted Radiotelephone Operator Permit must be at least fourteen (14).
(4) Each radio operator application form inquires as to the applicant's criminal record, if any, the status of his citizenship, and his physical ability to perform the duties of a radio operator.

EXAMINATIONS

(5) If applying for a Radiotelegraph type license, must successfully pass the prescribed code test consisting of both transmitting and receiving the International Morse Code for a period of one minute without error. This test is computed counting 5 letters per word or group with punctuation and numerals counting as 2 letters. It may be written in either pencil or ink. Semiautomatic keys and typewriters may be used for the 25 WPM test if furnished by the applicant. The speed requirements are as follows:

Radiotelegraph Third Class Operator Permit	16 code groups per minute and 20 words per minute, plain language.
Radiotelegraph Second Class Operator License	16 code groups per minute and 20 words per minute, plain language.
Radiotelegraph First Class Operator License	20 code groups per minute and 25 words per minute, plain language.

(6) Must be able to transmit and receive spoken messages in English and successfully pass written examination elements as follows:

Restricted Radiotelephone Operator Permit	No written examination required.
Radiotelephone Third Class Operator Permit	Elements 1 and 2.
Radiotelephone Second Class Operator License	Elements 1, 2, and 3.
Radiotelephone First Class Operator License	Elements 1, 2, 3, and 4.
Radiotelegraph Third Class Operator Permit	Elements 1, 2, and 5.
Radiotelegraph Second Class Operator License	Elements 1, 2, 5, and 6.
Radiotelegraph First Class Operator License	Elements 1, 2, 5, and 6.

The Commission's examination elements consist of the following:
No. 1, Basic Law—Provisions of laws, treaties and regulations with which every operator should be familiar. (20 Questions, multiple choice type)
No. 2, Basic Operating Practice—Operating procedures and practices generally followed or required in communicating by radiotelephone stations. (20 Questions, multiple choice type)
No. 3, Basic Radiotelephone—Technical, legal and other matters applicable to operating radiotelephone stations other than broadcast. (100 Questions, multiple choice type)
No. 4, Advanced Radiotelephone—Advanced technical, legal and other matters particularly applicable to operating various classes of broadcast stations. (50 Questions, multiple choice type)
No. 5, Radiotelegraph Operating Practice—Radio operating procedures and practices generally followed or required in communicating by radiotelegraph stations primarily other than in the maritime mobile services of public correspondence. (50 Questions, multiplue choice type)
No. 6, Advanced Radiotelegraph—Technical, legal and other matters applicable to operating all classes of radiotelegraph stations including

maritime mobile services of public correspondence, message traffic routing and accounting, radio navigational aids, etc. (100 Questions)
No. 7, Aircraft Radiotelegraph—Special endorsement on Radiotelegraph First and Second Class Operator Licenses. Theory and practice in operation of radio communication and navigational systems in use on aircraft. (100 Questions, multiple choice type; code test of 20 code groups per minute and 25 WPM plain language.)
No. 8, Ship Radar Techniques—Special endorsement on Radiotelegraph or Radiotelephone First or Second Class Operator Licenses. Specialized theory and practice applicable to proper installation, servicing and maintenance of ship radar equipment in use for marine navigational purposes. (50 Questions, multiple choice type)
No. 9, Basic Broadcast—Specialized elementary theory and practice in operation of standard (a-m) and fm broadcast stations. (20 Questions, multiple choice type)

(7) Except insofar as the requirement of one-year service for eligibility for Radiotelegraph First Class Operator License, as outlined under Item 12 herein, may be considered a training requirement, there are no educational or training requirements set up by the Commission as a prerequisite to taking an examination.

(8) Examinations for commercial radio operator licenses are conducted at each radio district office of the Commission on the days designated by the Engineer in Charge of the office. In addition to the radio district offices of the Commission, examinations are held in certain other cities on dates designated by the Engineer in Charge of the radio district in which these cities are located. A list of designated examination points will be forwarded upon request or when necessary to answer inquiries regarding such points. Specific dates and times of examinations should be obtained from the Engineer in Charge of the office concerned in each instance. *Available facilities do not permit extension of the regular radio operator license examination procedure to applicants overseas.* It is suggested that applicants overseas arrange for examination when they are able to appear at one of the Commission's designated examination points.

(9) The holder of a license, who applies for another class of license or special endorsement, will be required to pass only the additional written examination elements for the new class of license. Applicants should bring with them and present any licenses, permits and verification cards they may hold to the examiner at the time of examination. If the holder of a license qualifies for a higher class license in the same group, the license held will be submitted for cancellation and returned to the licensee upon issuance of the new license. Since code tests are not considered as "elements," credit for them is not generally allowed and it is necessary to re-qualify.

(10) An applicant who fails an examination element will be ineligible for a period of two months to take an examination for any class of license requiring that element. Examination elements will be graded in the order listed (not necessarily the same day completed), and an applicant may, without further application, be issued the class of license or permit for which he qualifies. *Seventy-five percent* is the passing grade for written examination elements.

(11) Any person who obtains or attempts to obtain, or assists another to obtain or attempt to obtain an operator license or permit by fraudulent means is committing a Federal offense for which severe penalties may be imposed.

PRIOR EXPERIENCE REQUIRED

(12) An applicant for a radiotelegraph first-class operator license shall have had an aggregate of one year of satisfactory service as an operator manipulating the key of a manually operated public ship or coast station handling public correspondence by radio telegraphy.

(13) To obtain employment as the sole radio operator on most cargo ships it is required that the licensed operator must have had at least six months prior satisfactory service as a qualified radiotelegraph operator in a station on board a ship or ships of the United States.

FEE SCHEDULE

(14) The following application filing fees must accompany applications for radio operator licenses and permits: $4 for first class, $4 for second class, $4 for third class, and $2 for commercial operator license endorsements, duplicates, renewals, and replacements, and $4 for restricted radiotelephone operator permits. Whenever an applicant requests both an operator license and an endorsement, the required fee will be the fee prescribed for the license document involved. Fees should be paid by check or money order payable to the Federal Communications Commission, *If an examination is to be taken at a place away from a field office, the application and fee should be filed in advance at the field office administering the examination.*

VERIFICATION OF LICENSE HOLDING

(15) Operators holding a radio operator license of the diploma form (other than Restricted Radiotelephone Operator Permit) may obtain a Verification Card, FCC Form 758-F, attesting to license holding. Verification Cards may be carried on the person of the operator in lieu of the license when operating a station at which the posting of an operator license is not required. When such Verification Cards are used the original license or permit must be readily accessible for inspection by an authorized government representative.

(16) If an operator is required to post his radio operator license at more than one station he may post the original license or permit at one station and post Certified Statements, FCC Form 759, at the other stations.

(17) Verification Cards or Verified Statements may be obtained by filing a properly completed application Form 756. The license or permit must accompany the request for verification. In lieu of the license or permit the operator shall exhibit a signed copy of the application which has been submitted by him until action is taken on the request.

RENEWALS, DUPLICATES, AND REPLACEMENTS

(18) An operator whose license or permit of the diploma form (other than Restricted Radiotelephone Operator Permits) has been lost,

mutilated or destroyed, shall immediately notify the Commission. An application Form 756 for a duplicate may be submitted to the *office issuing the original license or permit* embodying a statement attesting to the facts thereof. A replacement Restricted Radiotelephone Operator Permit may be requested by filing application Form 753 and the required fee with the Commission's Gettysburg, Pa. office.

(19) The holder of any license or permit whose name is legally changed may make application for a replacement document to indicate the new legal name, by submitting a properly completed application and the required fee to the office of original issue, accompanied by the license or permit affected.

(20) Licenses are normally renewable at any time within the last year of the license term or during a one-year period of grace after the date of expiration. During this grace period, an expired license is not valid. At this time it is not necessary to show service under a license sought to be renewed. (Renewal applications, when accompanied by the expiring license, should be filed at the nearest District Office. If the expiring license has been lost, destroyed, or mutilated, the application should be filed at the office which issued the original document. Applications may be by mail and should be submitted through the use of FCC Form 756.)

(21) When a duplicate or replacement operator license or permit has been requested or request has been made for renewal, or for an endorsement, in lieu of the license or permit, the operator shall exhibit a signed copy of the application which has been submitted by him.

GENERAL INFORMATION

(22) All licenses and permits other than amateur are considered to be commercial licenses and permits. Radio operator license requirements are usually governed by the type of emission involved and whether or not the operator's duties include making adjustments to transmitters.

(23) A license is not required for the operation of or repairs to radio or television *receiving* equipment.

(24) The Commission does not issue licenses for radio engineers, television engineers, television cameramen, radio mechanics, radio announcers or studio console operators. Persons who are employed at these jobs are required to hold operator licenses of the proper type and class issued by the Commission only if their duties include the operation of radio transmitting and/or video transmitting equipment.

(25) Restricted radiotelephone operator permits are normally issued for the lifetime of the holder. Commercial operator licenses and permits of other classes are normally issued for a term of five years from the date of issue.

(26) The minimum requirement of persons wishing to obtain employment as an operator at a standard (a-m) or fm broadcast station is a Radiotelephone Third-Class Operator Permit with the Basic Broadcast endorsement. (Examination Elements 1, 2, and 9). A nonrenewable Provisional Radio Operator Certificate, which carries with it all the authority now embraced by a Radiotelephone Third-

Class Operator Permit endorsed for Broadcast use, may be obtained by mail and without examination. The holder is expected to fulfill the examination requirements within the one year term of the certificate. Application may be made on FCC Form 756-C, which may be obtained from any of the Commission's Field Engineering Offices.

(27) Holders of Restricted Radiotelephone Operator Permits, Radiotelephone and Radiotelegraph Third-Class Operator Permits are in general prohibited from making adjustments that may result in improper transmitter operation. The Commission's Rules require that radio transmitting equipment operated by holders of these operator permits shall be so designed that none of the operations necessary to be performed during the course or normal rendition of service may cause off-frequency operation or result in any unauthorized radiation. Any needed adjustments to transmitters operated by holders of the Restricted Radiotelephone Operator Permit and Radiotelephone and Radiotelegraph Third-Class Operator Permits should be made by or in the presence of the holder of a higher class license of the proper grade.

(28) In general anyone wishing to obtain employment as an operator at a ship radiotelegraph station should hold a Radiotelegraph First- or Second-Class Operator License. Restricted Radiotelephone Operator Permits are valid for the normal operation of radiotelephone equipment installed in most aircraft, at certain ground stations, and on most boats where radio equipment is not mandatory.

(29) The class of commercial radio operator license which is normally required as sufficient authority to install, service, and maintain radiotelephone transmitting equipment on board aircraft and small boats and most radio transmitting equipment in the land-mobile services is the Radiotelephone Second-Class Operator License. Successful completion of Commercial Examination Elements 1, 2, and 3 is required prior to the issuance of this class of license.

APPENDIX VI

Answers to Self-Tests

ELEMENTS I AND II SELF-TEST

1. A.
2. B. If the original is found meanwhile it is necessary that you return either one, original or duplicate, to the FCC.
3. B.
4. D. Station logs must be signed when the operator goes on and off duty. Likewise log records must be signed by the licensed maintenance technician and must also include data on the nature of the transmitter repairs.
5. D.
6. B.
7. B.
8. A.
9. B.
10. D.
11. D.
12. A.
13. C.
14. A.
15. C.
16. B.
17. A. The operator must learn to be patient under adverse propagation conditions. Repeated calls should be made only if one is certain the transmissions cause no interference or hamper other communications that share the same frequency.
18. D.
19. C.
20. C.
21. A.
22. B.
23. C.
24. D.
25. B.
26. D.
27. D. Questions 26 and 27 show a method of questioning that is often used in FCC examinations. You are to select the most complete answer. In Question 27, both A and B are correct but the proper answer is D in terms of multiple-choice evaluation.
28. C.
29. D. Overmodulation is to be avoided because it limits the effective transmission range of the station because of speech distortion. However, it is not direct grounds for license suspension.
30. C.
31. B.
32. B.
33. B. This is a problem when one must transmit from a location near operating machinery or in

199

traffic. The development of noise-cancelling microphones has aided this problem. They must be close-talked to be really useful. Close-talk but do not shout.
34. A.
35. A.
36. D.
37. B.
38. D.
39. A.
40. A.

ELEMENT IX SELF-TEST

1. B. One kilohertz equals one thousand hertz. In one million hertz there are one thousand kilohertz.
2. C. The remote antenna meter reading is not a compulsory one and operation can continue. However, you should always notify the chief engineer or the first-class radiotelephone opertor in charge.
3. A.
4. C. Too high an audio level can cause overmodulation. If a decrease in the volume level setting does not overcome the problem it is your responsibility to notify the chief engineer or the first-class radiotelephone operator in charge.
5. D.
6. D.
7. D.
8. C.
9. D. You should strike out the erroneous portion (not erase) and initial the correction made and the date of correction.
10. A.
11. B. In this case there is no need to take the station off the air. Again it is the responsibility of the operator to notify the chief engineer or the first-class radiotelephone operator in charge.
12. C. Minor readjustments are always made after the official logging.
13. C.
14. A.
15. B. One kilovolt equals one thousand volts, therefore two thousand volts is the same as two kilovolts.
16. B.
17. B. 10% of the 1800-volt reading is 180 volts. Reading would be 1620 volts (1800 − 180).
18. C.
19. D. The present reading is 1.7, and 5% of this figure is 0.085. This would be an increase to 1.785.
20. A. In fact, failure of the remote control system would usually shut down the transmitter automatically.
21. D.
22. A.
23. C.
24. D. The limits are 5% above assigned power and 10% below assigned power.
25. A.
26. C. Abbreviations may be used if a list of such abbreviations is kept at some other point in the log.
27. B.
28. D.
29. B. This is not a condition for shutting down the station. However, it is your responsibility to notify immediately the chief engineer or the radiotelephone operator in charge.
30. A.
31. A.
32. C.
33. D.
34. This is the condition that exists after the transmission of an emergency action notification and before the transmission of the emergency action termination.
35. All other broadcast stations will observe broadcast silence.
36. D.
37. C.
38. A.
39. C.
40. C.
41. D.

42. A.
43. B.
44. A.
45. See R.R. 73.14(m).
46. Antenna resistance means the total resistance of the transmitting antenna system at the operating frequency and at the point at which the antenna current is measured.
47. Antenna input power or antenna power means the product of the square of the antenna current and the antenna resistance at the point where the current is measured.
48. The term rebroadcast means reception by radio of the program of a radio station, and the simultaneous or subsequent retransmission of such program by a broadcast station.
49. D.
50. C.
51. C. (1500 × 0.35)
52. A.
53. B.
54. B. 1.5/2.5
55. C.
56. B.
57. D.
58. C.
59. C.
60. D.

Index

A

Adjustments, transmitter, 25
Advisory, frequency, 40
Aeronautical
  -advisory stations, 34-35
  enroute stations, 32, 34
  fixed stations, 32, 34
  ground stations, 34-35
  metropolitan stations, 32, 34
  multicom stations, 32, 34-35
  public service airborne stations, 33
  radiocommunication, 35-37
  radionavigation, 35-37
  search and rescue mobile stations, 32, 34
  utility stations, 32, 34
Air carrier airborne stations, 33
Airborne stations, 33
Aircraft
  distress message for, 20
  radio installation, 36-37
Airdrome
  control stations, 34
  stations, 32
Alphabet, phonetic, 111
Ammeter, antenna, 72-73, 124-125
Amperes, 87
Antenna
  base-current ratio, 73
  current, 129
  directional, 73-74
    monitor for, 84-86
  power, 129, 130
  requirements, 62-63
  resistance, 129

APCO, 40
Associated Public-Safety Communications Officers, Inc., 40
A3H emission, 31
A3J emission, 31
Authorities, license, 9-13
Automobile emergency radio service, 55-56
Aviation radio services, 32-39

B

Base-current ratio, 73
Broadcast
  day, 87
  endorsement, 8, 69
Business radio service, 44, 49

C

Calling frequency, 26, 27, 65, 118
Carrier power, maximum rated, 129
Centers, radio-control, 40-42
Changeover, DSB to SSB, 29-30
Channels
  Class-D, 61-62
  ship, recommended, 27-28
Charges, message, 117-118
Citizens
  band equipment, 63-67
  Radio Service, 57-67
Civil Air Patrol stations, 32
Class
  -A stations, 61
  -C station, 61
  -D stations, 61-63
  I coast station, 22

Class—cont
  II coast station, 22
  III coast station, 22
Coast station
  Class I, 22
  Class II, 22
  Class III, 22
  limited, 22
  marine-utility, 22
  public, 22
Communications
  identification of, 17
  nature of, 16
  permissible, 59
  priority of, 16
  prohibited, 59-60
Control point, 47
Corrections, log, 17
Current
  antenna, 129
  plate, 71

D

Day, broadcast, 87
Daytime, 87
Definitions, 86-87
  technical, 129-130
Direct method, 74
Directional antennas, 73-74
  monitor for, 84-86
Distress, 37-38
  frequency, 26, 27
  message, 113-115
    for aircraft, 20
  procedure, 19-20, 24
  signal, 19, 115-116
Double-sideband transmission, 28-29
Duplex operation, 46
Duplicate operator permits, 92
Duration of transmissions, 60

E

EBS, 127
  test transmission, 91
Electrical terms, 87
Emergency, 37-38
  Action
    Notification, 91
    Termination, 91
  Broadcast System, 14, 91
  frequency, 65
  operation, 112-113
Endorsement, broadcast, 8, 69

Equipment
  citizens band, 63-67
  remote control, 88
  tests, 38
Examination, FCC, 95
  requirements, 98

F

FCC, 93
  district offices, 93
  forms, 93
  Rules and Regulations, 86-92
Federal Communications
    Commission, 93
Fees
  payment of, 102
  schedule of, 102-103
Field, effective, 129
Fines, 18
Fire service, 39
Fixed
  relay station, 44
  station, 44
Flight test
  and flying school airborne
    stations, 33
  stations, 32
Flying school stations, 32
Fm
  broadcast
    band, 87
    station, 87
  stereo, 88-89
  stereophonic broadcast, 87
Forest radio service, 44, 48
Forestry-conservation service, 39
Forms, FCC, 32, 58
Frequency
  advisory, 40
  bands, 45-46
  calling, 26, 27, 65, 118
  distress, 26, 27
  emergency, 65
  intership, 27
  tolerance, 70, 125
  working, 27, 118

G

Grace period, 92

H

Harmonics, audio, combined, 129
Highway service, 39

I

Identification
  communications, of, 17
  sponsor, 89-90
  station, 61, 38-39, 47, 89, 110, 126
Imprisonment, 18
IMTS telephone system, 53-55
Indirect method, 74
Industrial radio
  location service, 44
  services, 42-55
Inspection
  station, 88
  US vessels by other countries, 117
Inspector, FCC, information
    available for, 126
Interference, 109
  harmful, 107
  prevention of, 17
Intership frequencies, 27

K

Kilovolt, 87
Kilowatt, 87

L

Land Transportation Radio
    Services, 55-56
License
  authorities, 9-13
  considerations, 8-14
  lost, 105
  operator, 8, 15-16
  renewal of, 106
  station, 8, 15
  term, 105
Licensing program, FCC, 69
Lighting, tower, 91-92, 111
Limited
  coast stations, 22
  ship station, 22
Local government service, 39
Log, 26, 126
  operating, 78-79
  program, 77
  radio, 17
  requirements, 76-79
  station, 106
Logging, corrections after, 127
Lotteries, 90

M

Manufacturers radio service, 44, 48

Marine
  radiotelephone, 28-32
  -utility coast stations, 22
Maritime radio services, 21-32
Marking, tower, 111
"Mayday," 19
Message, contents of, 107
Meter
  modulation, 72
  power-output, 72
Metering, transmitter, 71-74
Microphone, use of, 110
Milliampere, 87
Mobile
  -base station, 44
  relay station, 44
Modulation, 70
  excessive, 72
  meter, 72
  monitor, defective, 126-127
  percentage of, 125, 129-130
Monitor
  directional antenna, 84-86
  fm, 82
Monitoring, 47
Morse code, 47
Motion picture radio service, 44, 50-51
Motor carrier radio service, 55, 56

N

Nighttime, 87
Nominal power, 87, 129
Notice of violations, 18

O

156.8 megahertz, 27
Operating
  authority, 98-100
  log, 78-79
  power, 74, 129
  procedures, 25
Operation, remote control, 88
Operational stations, 32, 34
Operator
  license, 8, 15-16
  permits, 15-16
    duplicate, 92
    radiotelephone
      restricted, 32
    third-class, 9-12, 98-99
  renewed, 92

Operator—cont
 permits
  restricted radiotelephone, 98, 99-100
  third-class, 8
 radio, essential provisions for, 15-20
 requirements, 9, 23, 33, 39, 46-47, 98-103
 responsibilities, 71
 third-class, stations operated by, 125
Overmodulation, 70

P

"Pan," 19
Penalties, 18-19, 107
Permits, operator, 15-16
 duplicate, 92
 posting of, 88
 radiotelephone
  restricted, 8, 9, 12-13, 32, 98, 99-100
  third class, 98-99
 renewed, 92
 third-class, 8
Petroleum radio service, 44, 48
Phonetic alphabet, 111
Plate
 current, 71
 input power, 129
 voltage, 71
Police service, 39
Posting of operator permits, 88
Power
 antenna, 129
  input, 130
 carrier, maximum rated, 129
 direct method, 74
 indirect method, 74
 limits, 70, 125
 nominal, 87, 129
 operating, 74, 129
 -output meter, 72
 plate input, 129
 radio service, 44, 47-48
Private airborne stations, 33
Privileges, special, 100-102
Profanity, 112
Program log, 77
Public
 coast stations, 22
 correspondence, 22
 Safety Radio Services, 39-42

Public—cont
 ship station, 22

R

Radio
 compulsory on ships, 21-22
 -control centers, 40-42
 installation, aircraft, 36-37
 log, 17
 operators, essential provisions for, 15-20
 services, aviation, 32-39
Radiocommunication
 aeronautical, 35-37
 secrecy of, 16
Radionavigation
 aeronautical, 35-37
 land stations, 32, 34
Radiotelephone
 marine, 28-32
 operator permit, restricted, 8, 9, 12-13, 32, 98, 99-100
 rule reminders, ship, 23-24
 third-class operator permit, 9-12, 98-99
Railroad radio service, 55, 56
Ratio, base-current, 73
Rebroadcast, 89, 106
Recorded material, 89
 announcement as, 127
Records, station, availability of, 88
Reexamination, eligibility for, 98
Relay press radio service, 44, 51
Remote control
 equipment, 88
 failure of, 125
 operation, 81-84, 88
Renewed operator permits, 92
Requirements, operator, 9
Resistance, antenna, 129
Restricted radiotelephone operator permit, 8, 9, 12-13, 32, 98, 99-100
Restrictions on use of Citizens band, 58-61
Rules and Regulations, FCC, 79, 86-92
 subscription to, 94-95

S

Safety
 message, 113-114
 signal, 19, 115-116

SCA, 77, 87
Secrecy of radiocommunications, 16
"Security," 19
Service, radio
  automobile emergency, 55-56
  business, 44, 49
  Citizens, 57-67
  fire, 39
  forest, 44, 48
  forestry-conservation, 39
  highway, 39
  industrial, 42-55
    radio location, 44
  land transportation, 55-56
  local government, 39
  manufacturers, 44, 48
  maritime, 21-32
  motion picture, 44, 50-51
  motor carrier, 55, 56
  petroleum, 44, 48
  police, 39
  power, 44, 47-48
  public safety, 39-42
  railroad, 55, 56
  relay press, 44, 51
  special
    emergency, 39
    industrial, 44, 49
  state guard, 39
  taxicab, 55
  telephone-maintenance, 44, 51-53
Ship
  radiotelephone rule reminders, 23-24
  stations, 22
    channels recommended for, 27-28
    limited, 22
    public, 22
Signal
  distress, 19
  safety, 19
  urgency, 19
Simplex operation, 46
Single sideband
  changeover to, 29-30
  transmission, 28-29
Special
  emergency service, 39
  industrial radio service, 44, 49
  privileges, 13-14, 75-76, 100-102
Sponsor
  identification, 89-90
  name, 127
SSB; *see* single sideband

Standard broadcast
  band, 86
  station, 86
State guard service, 39
Station(s)
  aeronautical
    advisory, 34-35
    enroute, 32, 34
    fixed, 32, 34
    ground, 34-35
    metropolitan, 32, 34
    multicom, 32, 34-35
    search and rescue mobile, 32, 34
    utility, 32, 34
  airborne
    aeronautical
      advisory, 32
      public service, 33
    air carrier, 33
    flight test and flying school, 33
    private, 33
  airdrome, 32
    control, 34
  categories, 47-53, 55-56, 61-63
  Civil Air Patrol, 32
  fixed, 44
    relay, 44
  flight test, 32
  flying school, 32
  fm broadcast, 87
  identification, 38-39, 47, 61, 89, 110, 126
  inspection, 88
  license, 8, 15
  mobile
    -base, 44
    relay, 44
  operation, 38
  operational, 32, 34
  plans, 79-86
  radionavigation land, 32, 34
  records, availability of, 88
  requirements, 75-76
  ship, 22
  standard broadcast, 86
  VOR, 35-36
Stereo, fm, 88-89
Stereophonic broadcast, fm, 87
Subsidiary Communications Authorization, fm, 87, 88-89
Sunrise, 87
Sunset, 87
Suspension, license, 106-107
  notice of, 107

T

Taped material, 89
Taxicab radio service, 55
Technical considerations, 70-71
Telegraphy, 13
Telephone
    conversations, broadcast of, 90-91
    -maintenance radio service, 44, 51-53
    system, IMTS, 53-55
Test(s)
    EBS, 128
    equipment, 38
    transmissions, 116-117
    system, 91
Testing of transmitter, 111
Third-class
    operator
        permit, 8
        stations operated by, 125
    radiotelephone operator permit, 9-12, 98-99
Tower
    lighting, 91-92, 111, 127
        light-sensitive device, control by, 128

Tower—cont
    marking, 111
Transmissions, duration of, 60
Transmitter(s), 26
    adjustments, 25
    metering, 71-74
    unattended, 112
2182 kilohertz, 26-27, 113

U

Urgency signal, 19, 115-116

V

Vessels, radio compulsory on, 21-22
Violations, notice of, 18, 107
Voice, training of, 110
Voltage, plate, 71
Volts, 87
VOR stations, 32, 35-36

W

Watch on 2182 kHz, 115
Watts, 87
Words, procedure, 110-111
Working frequencies, 27, 118